STATE OF THE ENVIRONMENT REPORT ON THE ST. LAWRENCE RIVER

VOLUME 2

The State of the St. Lawrence

Environment Canada
Environnement Canada
Quebec Region
Région du Québec

Canadian Cataloguing in Publication Data

Main entry under title:

State of the Environment Report on the St. Lawrence River

Issued also in French under title: Rapport-synthèse sur l'état du Saint-Laurent.

To be complete in 2 v. and a brochure.

Includes bibliographical references.

Summary : v. 1. The St. Lawrence Ecosystem – v. 2. The State of the St. Lawrence.

Co-published by: St. Lawrence Centre.

ISBN 2-921146-34-7 (set) – ISBN 2-921146-31-2 (v. 1) – ISBN 0-660-16382-9 (v. 1) (St. Lawrence Centre) – ISBN 2-921146-32-0 (v. 2) – ISBN 0-660-16385-3 (v. 2) (St. Lawrence Centre)

1. Ecology – St. Lawrence River. 2. Biotic communities – St. Lawrence River. 3. Stream ecology – St. Lawrence River. 4. Watersheds – St. Lawrence River. I. Lavigne, Nicole. II. Bouchard, Hélène. III. Provencher, Michel-A. IV. St. Lawrence Centre.

QH106.2.S35R3613 1996 574.5'26323'09714 C96-940299-6

Cover page photograph: Michel Boulianne

ISBN 2-921146-34-7 (set) Éditions MultiMondes

ISBN 2-921146-32-0 (Vol. 2) Éditions MultiMondes

ISBN 0-660-16385-3 St. Lawrence Centre

Cat. No.: En153-70/2-1996E

Legal Deposit – Bibliothèque nationale du Québec, 1996

Legal Deposit – National Library of Canada, 1996

Published by Éditions MultiMondes Inc. and Environment Canada, St. Lawrence Centre, in collaboration with the Canada Communication Group – Publishing, Supply and Services Canada.

Éditions MultiMondes
930 Pouliot Street
Sainte-Foy, Quebec
Canada G1V 3N9
Tel.: (418) 651-3885
Fax: (418) 651-6822
1 800 840-3029
E-Mail: info@multim.com
Internet: http://multim.com

St. Lawrence Centre
Environmental Conservation
Quebec Region
Environment Canada
105 McGill Street, 4th Floor
Montreal, Quebec
Canada H2Y 2E7
Tel.: (514) 283-7000

The inside pages are printed on paper made up of 50% recycled fibre and 10% of post-consumption recycled fibre. The cover page is printed of cardboard made up of 40% recycled fibre and 20% post-consumption recycled fibre.

Notice to Readers: The St. Lawrence Centre thanks all those who participated in the production of this report and assumes full responsibility for its contents.

To be cited as: St. Lawrence Centre. 1996. *State of the Environment Report on the St. Lawrence River. Volume 2: The State of the St. Lawrence.* Environment Canada – Quebec Region, Environmental Conservation, and Éditions MultiMondes, Montreal. "St. Lawrence UPDATE" series.

PRODUCTION

The State of the St. Lawrence

ENVIRONMENT CANADA – QUEBEC REGION

St. Lawrence Centre

Design and Structure:	Nicole Lavigne Hélène Bouchard Michel-A. Provencher
Co-ordination:	Hélène Bouchard
Research and Writing:	Hélène Bouchard Danielle Gingras Louise Quilliam Claudine Loiselle

CONSORTIUM GAUTHIER & GUILLEMETTE CONSULTANTS AND COMMUNICATIONS SCIENCE-IMPACT

Science Consultant:	Jean-Hugues Bédard
Special Contributors:	Anne Bédard Edwin Bourget Pierre Fréchette Guy Lacroix Jean Lebel Serge Lepage André Nadeau Sylvie Trépanier
Science Writer:	Jean-Marc Gagnon
Project Manager:	René Nault
Project Co-ordinators:	Anne Bédard Sylvie Trépanier

The Consortium would also like to thank the following persons, who made suggestions, provided uncited information or offered pertinent comments throughout the various stages of production of this report: Pierre Bertrand, Dan Busby, François Caron, Gilles Chapdelaine, Louis Désilets, Denis Fournier, Élyse Lauzon, Yves Lavergne, Serge Pilote, Austin Reed, Marcel Shields, Jean Therrien, Serge Tremblay.

ADDITIONAL CONSULTANT

Writer:	Mariette Tremblay

ADVISORY COMMITTEE

Environment Canada:	Marie-José Auclair Caroll Bélanger Alain Bernier Jean Burton Lynn Cleary Thérèse Drapeau Clément Dugas Annie-France Gravel Lucie Olivier Claude St-Charles Aline Sylvestre Jacqueline Vincent
Fisheries and Oceans:	Louis Désilets Jean-François La Rue
St. Lawrence Action Team:	Marc Villeneuve
Ministère de l'Environnement et de la Faune:	Fay Cotton Roger Lemire Raymond Martin

ENVIRONMENT CANADA – QUEBEC REGION

St. Lawrence Centre

Linguistic Revision:	Patricia Potvin
Graphics:	Denise Séguin
Cartography:	François Boudreault
Production Follow-up:	Louise Quilliam Denise Séguin Hélène Bouchard
Photographic Research:	Danielle Gingras

PUBLIC WORKS AND GOVERNMENT SERVICES CANADA – QUEBEC REGION

Translation Bureau

Translation:	Montreal English Translation Section

ÉDITIONS MULTIMONDES

Graphic Design:	Norman Dupuis
Graphics:	Mathieu Gagnon Gérard Beaudry

CONSULTANTS

Photoengraving:	Compélec Inc.
Printing:	Imprimerie La Renaissance

ACKNOWLEDGMENTS

We would like to thank the undermentioned for their help at the various stages of production of this report:

at the Canadian Coast Guard, Micheline Moore,

at Environment Canada, Alain Armellin, Thérèse Baribeau, Jean-Charles Bellemare, Luc Bergeron, Gilles Chapdelaine, Josée Deguise, Yves de Lafontaine, Karima Fazal, Louise Lapierre, Guy Létourneau, Jean-René Michaud, Louise Quilliam, Isabelle Ringuet, René Rochon, Raymond Van Coillie, Raymonde Verville,

at Fisheries and Oceans, Éric Gagnon, Louise Savard, Yvan Vigneault,

at the St. Lawrence Action Team, Gaétan Duchesneau, Gilles Legault,

at the Ministère de l'Environnement et de la Faune, Daniel Banville, Marcel Bernard, Pierre Bilodeau, François Caron, Richard Chatelain, Denyse Gouin, Yves Grimard, Claude Grondin, Serge Hébert, Michel Huot, Rosaire Jean, Marcel Lacasse, René Lafond, Denis Laliberté, Denis Lapointe, Jean-Pierre Le Bel, Michel Lepage, Camille Paré, Francine Richard, Louis Roy, Serge Tremblay, Guy Trencia, Guy Verreault,

at Stratégies Saint-Laurent, Pierre Lemay.

PREFACE

The St. Lawrence Action Plan was launched in 1988 by the governments of Canada and Quebec to restore, protect and conserve the St. Lawrence River environment.

The plan had the following objectives:

1. To reduce by 90% the liquid toxic waste discharged by 50 priority industrial plants by 1993.
2. To implement plans to restore contaminated federal sites and wetlands.
3. To protect 5000 hectares of wildlife habitats and create a marine park at the mouth of the Saguenay River.
4. To develop and implement recovery plans for certain threatened species.
5. To produce a comprehensive report on the state of the St. Lawrence River environment.

The fifth objective, to produce a comprehensive report on the state of the St. Lawrence, is one of the principal mandates of the St. Lawrence Centre (SLC).

The purpose of the report is to review current knowledge about the resources and uses of the St. Lawrence, the pressures to which it is subjected, and its limitations and possibilities. In an effort to make such information widely available, the SLC has produced different documents for a varied audience. These include such materials as the *Environmental Atlas of the St. Lawrence* (11 plates so far), fact sheets and reports. The approach to producing the comprehensive report is distinguished by its modular and progressive nature: different kinds of documents (or tools) are prepared for different target groups as environmental data becomes available.

A key component of the comprehensive report is the state of the environment (SOE) report, which integrates and analyses the latest scientific findings to arrive at an objective assessment of the state of the St. Lawrence ecosystem. The process is like preparing a financial statement, but in an extremely complex context: an ecosystem which has almost unlimited variables (to date identified only in part) and, even more importantly, which is made up of components so closely interwoven that useful conclusions can often be obtained only by extrapolation. Note that, in the interests of objectivity, the report makes no attempt to evaluate any programs targeting the St. Lawrence. The SOE report presents findings on the state of the St. Lawrence ecosystem; its purpose is to help communities and decision makers make informed decisions about the environment.

The SOE report is designed to meet the needs of different groups. It is divided into two volumes. The first volume, *The St. Lawrence Ecosystem*, describes the physico-chemical, biological and socio-economic state of the St. Lawrence and presents the methodology used to incorporate the scientific findings on which the assessment is based. It is aimed at an informed public composed mainly of scientists, teachers, environmental groups, government departments and consultants interested in a reference work on various aspects of the St. Lawrence. The second volume, *The State of the St. Lawrence*, uses the latest scientific findings to diagnose the state of the St. Lawrence ecosystem. It was designed to give decision makers greater access to scientific data and to help them make knowledgeable decisions about the St. Lawrence.

Lastly, a major concern during the preparation of the state of the environment report was to encourage the public to make active use of the information provided. The report seeks to respond to the questions and expectations of the general public. This was also the intent behind a document for non-specialists that summarizes the main conclusions of the SOE report.

FOREWORD

Preparing a profile of the state of the St. Lawrence is a highly complex operation because synthesizing data to obtain a simple global assessment of the health of the St. Lawrence is a major challenge. Data are so extensive and diverse, and are obtained from such different environments, that it is difficult to believe they are about the same waterway. In some places the St. Lawrence is a river, in others a lake; some parts of it are fresh water, others salt water; the only constant is the water itself. The characteristics and quality of the water and its links with the natural environments and living species are therefore key to all activities in the St. Lawrence environment.

Environmental assessments are important information sources. They usually include a sector-by-sector analysis of different aspects of the environment (such as fisheries, agriculture and forests) and sometimes even separate the activity (agriculture, for example) from its physical support (such as the soil). Such an approach provides an accurate picture of each environmental component, but leaves out the links between resources and human activity.

Employing an ecosystem approach to take these links into account, the challenges of the state of the environment report on the St. Lawrence were as follows:

1. Discuss environmental data in a text for the non-scientist without sacrificing scientific accuracy.

   Scientific knowledge is constantly being updated. Since statements about the state of the St. Lawrence had to based on available information, some gaps are inevitable.

2. Tangibly increase the use of environmental information in decision-making processes.

   Despite scientific uncertainty, we must develop a more global interpretation of data on the St. Lawrence ecosystem so that science has greater influence on decision making.

3. Help improve environmental monitoring of the St. Lawrence.

   It is essential that gaps in scientific knowledge be identified so that monitoring of the St. Lawrence can be more systematic.

The method used to assess the state of the St. Lawrence was based on an ecosystem approach. Using a matrix that represents and weights influences with the greatest impact on the state of the St. Lawrence, the state of the environment report presents a detailed assessment of the ecosystem. It also identifies the characteristics to be monitored and the gaps in knowledge that need to be filled in future years in order to measure changing conditions. The profile is rounded out by inclusion of state of the environment indicators for each characteristic.

The tools used in this report are the matrix and indicators; it goes without saying that they will probably be modified as new information becomes available. The goal of the report will be achieved when decision makers – be they individuals, teachers, businesses, environmental groups or governments – use the best possible information available in their decision-making process, and when a dynamic approach that keeps pace with scientific findings is developed to monitor the St. Lawrence ecosystem. The St. Lawrence Vision 2000 action plan calls for concerted efforts to fill the gaps in our knowledge. When that plan ends, a second joint report on the state of the St. Lawrence will be prepared by the Vision 2000 partners. The second report will focus on specific aspects of the St. Lawrence ecosystem and will select certain areas for more detailed analysis. This first report is therefore one stage in the synthesis and integration of current knowledge; it is a stepping stone in the continual process of exploring and applying analytical approaches and methods to our highly diverse knowledge of the St. Lawrence River.

TABLE OF CONTENTS

LIST OF FIGURES

LIST OF TABLES

LIST OF INFORMATION SUPPLEMENTS

Introduction

THE ST. LAWRENCE: A VITAL ARTERY

The St. Lawrence River is a natural environment of great richness. From the Great Lakes to the Atlantic Ocean it passes through very diverse habitats, widening into a number of shallow lakes, stretching into aquatic plant beds and marshes and flooding forests along its banks. It then turns into a vast estuary and forms a typically maritime region with unique oceanographic features, before finally merging into the waters of the Gulf.

Despite its northern latitude, the St. Lawrence contains a wide variety of habitats that shelter a wealth of plant and animal life. It is home to 185 species of fish and some 20 species of marine mammal; over 150 bird species inhabit its shorelines or use it as a staging area during migrations; about 1300 vascular plant species grow in its waters. The River is also a remarkable natural showcase: people come from near and far to watch migrating Greater snow geese at Baie-du-Febvre and Cap Tourmente, hear the rorquals spouting just off Escoumins or marvel at the majestic granite cliffs of the Saguenay Fjord.

THE ST. LAWRENCE: OUR GREATEST WATERWAY

Once a route for Huron and Iroquois canoes and small flat-bottomed French fishing boats, today the St. Lawrence is actively plied by bulk carriers, container ships and oil tankers, with more than 10 000 trips per year. Few rivers penetrate so deeply into the heartland of a continent and have such a powerful impact on a country's development.

Today, 60% of Quebec's inhabitants live in riverside towns and cities and the vast majority of industries and manufacturing companies are established along the riverbanks. In the mid-1980s, half the population of Quebec was fed by the fertile lands of the great plain crossed by the St. Lawrence.

THE ST. LAWRENCE: DRIVING OUR ECONOMY

The St. Lawrence is a tremendous resource. It provides water for many industrial and domestic uses – close to 46% of Quebecers draw their drinking water directly from it – and offers numerous biological resources. Even though the commercial freshwater fishery has declined considerably in the past forty years, it is still a significant economic activity. In 1992, the annual freshwater catch was 965 tonnes, consisting mainly of Brown bullhead, Yellow perch, Lake sturgeon and American eel. The commercial marine fishery in the estuary and Gulf is even larger: landings in 1992 – 70 413 tonnes – were worth $88.9 million. The fishery employs over 5000 people, mostly in the Gaspé.

An ever-increasing percentage of the population uses the St. Lawrence for recreation and tourism. The St. Lawrence has tremendous potential for such activities as swimming, sailboarding, kayaking, canoeing, sport fishing, waterfowl hunting and nature watching. Pleasure boating is also becoming more popular.

In the late 1950s, construction of the St. Lawrence Seaway greatly augmented shipping traffic. At the same time, new agricultural practices based on the massive use of chemicals and fertilizers became more common; this, along with industrialization and the growth of riverside populations, affected the ecological balance of the St. Lawrence. The state of the St. Lawrence is still a cause for concern today. Industrial effluents loaded with toxic substances, contaminated sediments, deformed fish and cancerous Beluga whales all point to the deterioration of the ecosystem and its biological resources.

Structure of the SOE Report

The State of the St. Lawrence was produced to present the conclusions of the *State of the Environment Report on the St. Lawrence River*, based on the latest figures available. It is designed as a management tool, structuring environmental information so as to assist decision making. This report is the fruit of many months work and discussion by the St. Lawrence Action Plan partners and a number of collaborators.

The report is divided into three chapters. Chapter 1 provides an overview of the fundamental features and particularities of the four hydrographic regions of the St. Lawrence.

Chapter 2 sets out the conceptual framework used to assess the state of the St. Lawrence and improve environmental monitoring. The framework is summarized in a chart that indicates how the St. Lawrence ecosystem works, using the 14 characteristics that are most indicative of the state of the St. Lawrence, combined with a set of indicators. The indicators are a kind of "barometer" of the health of the St. Lawrence and are particularly important for decision makers.

Chapter 3 discusses the general state of the St. Lawrence. After a background section on the conclusions of the first general report issued in the late 1970s, the state of the St. Lawrence is assessed on the basis of the 14 selected characteristics.

Lastly, the conclusion emphasizes the importance of environmental monitoring. Regular assessments of the St. Lawrence ecosystem would become a dynamic tool, providing invaluable information to those who make medium- and long-term decisions about the future of the St. Lawrence.

READER'S GUIDE

Refers the reader to a given section (4.2.2) in Part 4 (P4): Method of Integration and Main Results of Volume 1 (V1).

Refers the reader to appendix map 1.

1

Four Rivers In One

Hydrographic Regions of the River: Physico-Chemical, Biological and Socio-Economic Aspects

The Great Lakes–St. Lawrence system ranks thirteenth among the world's hydrographic networks for the size of its drainage basin (1.6 million km^2), sixteenth for its mean annual discharge of 12 600 m^3/s at Quebec City and seventeenth for its 3260-km length (from Lake Superior to the Cabot Strait). The system drains over 25% of the world's fresh water supply.

The hydrographic conditions, physical and chemical characteristics of the water, the aquatic plants, terrestrial plants and wildlife all change significantly along the length of the River. In fact, a system as long as the St. Lawrence could not possibly remain the same throughout. Thus, for the purposes of study and analysis, the St. Lawrence can be divided into four hydrographic and ecological sectors, each with its own characteristics. They are as follows:

1. **Fluvial Section** (from Cornwall to Pointe-du-Lac)
2. **Fluvial Estuary** (from Pointe-du-Lac to the eastern tip of Île d'Orléans)
3. **Upper Estuary** (from the eastern tip of Île d'Orléans to Tadoussac, and including the Saguenay River)
4. **Lower Estuary** (from Tadoussac to Pointe-des-Monts) and **Gulf**.

The hydrographic divisions adopted by the St. Lawrence Action Plan are shown in Figure 1.1. Note that, for the *State of the Environment Report on the St. Lawrence River*, the western limit of the Fluvial Section was set at Cornwall to correspond to the geographic area covered by the St. Lawrence Action Plan.

1.1 THE FLUVIAL SECTION

The southernmost part of the St. Lawrence ecosystem is characterized by rapids, lakes, islands, islets and a number of hydraulic structures used to generate

FIGURE 1.1

Hydrographic regions of the St. Lawrence

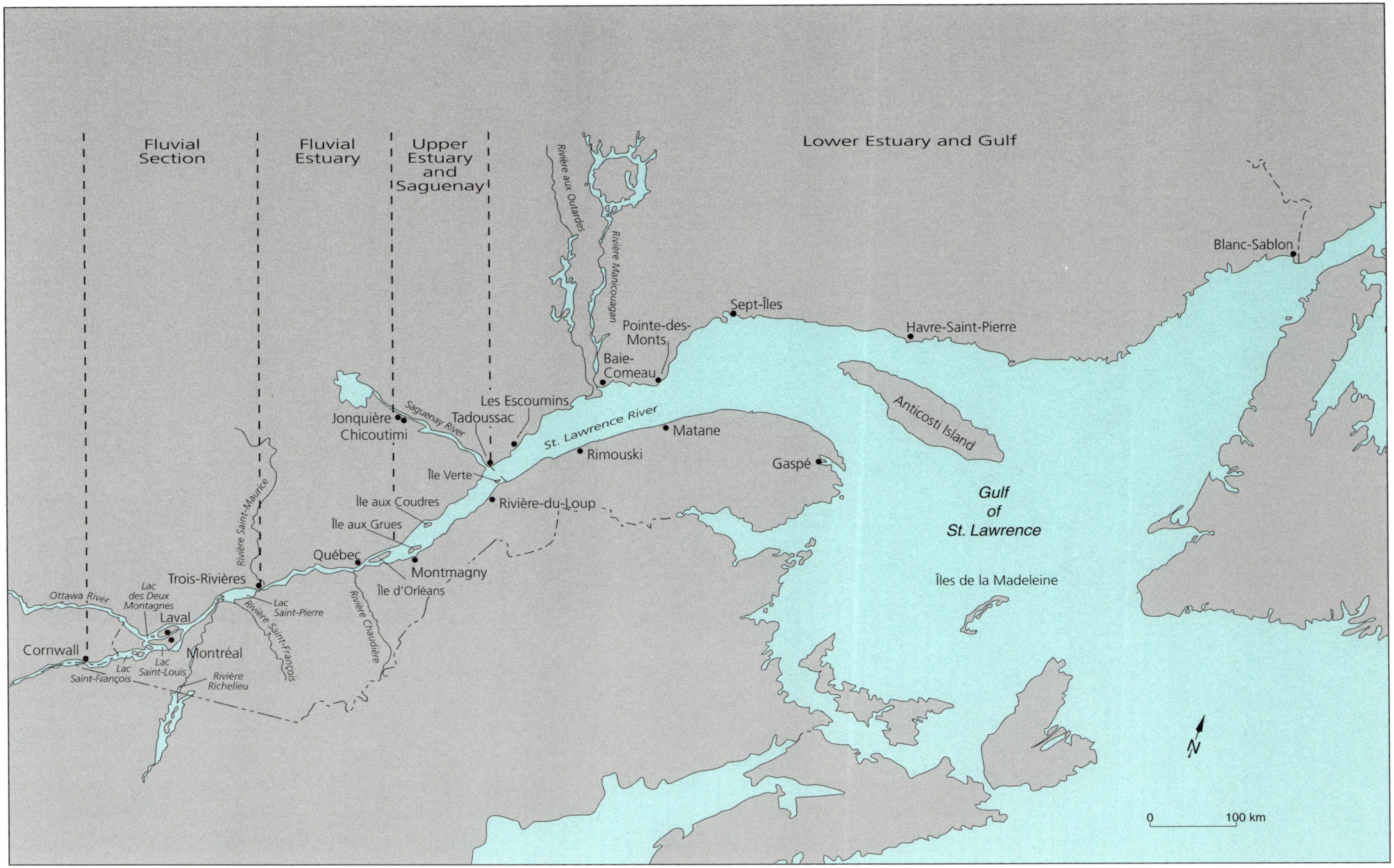

electricity and aid commercial shipping. The Fluvial Section, or main course of the River, also passes through the most fertile lands of the St. Lawrence system.

The freshwater stretch starts as a long corridor then broadens to form several large shallow basins, including Lake Saint-François and Lake Saint-Louis. Water in this stretch is regulated for hydro-electric production and to aid vessels of large tonnage. Its distinguishing feature is the St. Lawrence Seaway, a series of locks and dredged sections that allows ships to negotiate the 68-metre vertical rise between Lake Ontario and Lake Saint-Pierre – the downstream limit of this part of the St. Lawrence.

The Fluvial Section has its source in the Great Lakes. It drains the Ottawa, Châteauguay, Richelieu, Nicolet and Saint-François rivers, tributaries that contribute to fluctuations in the River's discharge. These tributary waters flow alongside the shoreline over varying distances before mixing into the water of the River.

Over one hundred islands change the flow of the St. Lawrence between Montreal and Sorel, trapping sediments. The weak currents in these sedimentation zones are choice habitats for fish. The sediments in these habitats are composed of the same kind of fine particles found in suspended sediments; this makes them highly susceptible to contaminant accumulation.

The Fluvial Section contains the largest area of shoreline wetlands. The richest area is Lake Saint-Pierre, which has about 80 fish species and some 60 aquatic plant species. Wetlands have highly diversified communities. Aquatic plants flourish in the shallower parts of the system, supporting varied populations of benthic invertebrates and fish. Depending on the area, plants form fairly exclusive communities with specific characteristics. Vegetation grows in bands or belts. Plant communities are dominated by submerged and emergent species; upland lie marshes, wet meadows and swamps with shrubs or trees, ideal habitats for waterfowl, wading birds and several of Quebec's rare bird species. The flood plain, which sometimes stretches beyond the swamps to farmlands, floods only in spring and plays an important role in the functioning of the ecosystem. In spring, a number of fish species live out part of their life cycle in this habitat.

Over three million inhabitants (75% of riverside communities) are concentrated in the 240-km section between Cornwall and Trois-Rivières, in the most urbanized zones. This stretch includes the island of Montreal, with a population density of 3600 inhabitants/km^2. Most drinking water intakes for these communities are located in the Fluvial Section of the River.

Several commercial ports are located along this section of the St. Lawrence, including the three main ports (i.e., those which annually handle over one million tonnes of goods) of Montreal, Contrecoeur and Sorel. This part of the St. Lawrence is also best equipped with pleasure-craft infrastructures.

The main course of the River, especially in the Lake Saint-Pierre region, has the best freshwater commercial fishing, sport fishing and waterfowl hunting. It is also the most highly industrialized region: in 1993, the area was home to 27 of the 50 industrial plants targeted under the St. Lawrence Action Plan in the pulp and paper, metallurgy, chemicals, and petrochemicals sectors, and in surface treatment and wood preservation industries.

1.2 THE FLUVIAL ESTUARY

From the outlet of Lake Saint-Pierre to the eastern tip of Île d'Orléans, the river bed narrows and straightens. The St. Lawrence now flows between two steep banks. The aquatic grass beds and marshes so abundant upstream disappear almost completely. Below Portneuf, they give way to rocky flats subject to tidal action and to ice abrasion in the spring, two factors that contribute greatly to sediment transport.

The waters of the St. Lawrence flow with no significant mixing down to Portneuf. In the Quebec City area, the reversal of the current that occurs with the rising tide causes greater mixing of the tributary waters, which come mainly from the Saint-Maurice, Batiscan, Sainte-Anne and Bécancour rivers.

The wetlands are composed of marshes, wet meadows and swamps and cover a much smaller area in this stretch of the St. Lawrence than upriver. Along with the Fluvial Section, the Fluvial Estuary is where the greatest number of different plant communities have been identified. The region, between the downstream section of the Fluvial Estuary and the upstream part of the Upper Estuary, probably contains the most diverse inventory of aquatic birds, which include the Greater snow goose and several species of duck. But, generally speaking, the diversity of animal communities decreases; this is particularly true of fish, which drop from 80 species in the Fluvial Section to 55 species in the Fluvial Estuary.

Two major urban centres, Trois-Rivières (50 000 inhabitants) and the Quebec Urban Community (500 000 inhabitants), are located in this 160-km stretch of the St. Lawrence.

Among the main commercial ports on the St. Lawrence are the ports of Quebec City, Bécancour and Trois-Rivières. For many years the port of Quebec City was the largest seaport in Canada.

There is much less commercial and sport fishing in the Fluvial Estuary than in the Fluvial Section. The main commercially-fished species in 1992 were American eel, Lake sturgeon and Yellow perch. The fertile lands bordering the Fluvial Estuary are mainly used for forage crops and intensive farming of animals such as hogs.

The manufacturing industry in Quebec City is fairly diversified, though smaller than in Montreal. The city's main economic base is government services. Ten of the fifty industrial plants targeted under the Action Plan are located in the Fluvial Estuary. These plants are in the pulp and paper, metallurgy, chemicals, petrochemicals and aluminum sectors.

1.3 THE UPPER ESTUARY AND SAGUENAY

The Upper Estuary extends from the eastern tip of Île d'Orléans to the mouth of the Saguenay River, forming a transition zone where fresh water from the River and its tributaries mixes with salt water from the Atlantic Ocean. More than 50 islands and islets are scattered throughout this section of the St. Lawrence. The largest ones are the Île aux Grues archipelago, Île aux Coudres, the Kamouraska islands, Île aux Lièvres and Île Verte. The main tributaries are the Saguenay, Malbaie and du Sud rivers. The Saguenay Fjord has several striking features, including troughs almost 300 metres deep and a sill along its mouth. Its endemic marine life, isolated within the geographic enclave, is characteristic of arctic regions.

A number of factors pose obstacles to the survival of freshwater organisms in the Upper Estuary. The increasing salinity below Île d'Orléans and the high turbidity up to Île aux Coudres, which contrasts with the relative clarity of much of the River, as well as the action of powerful physical forces – a large tidal range that continuously exposes and inundates the shoreline and the continual turbulence of water masses – make conditions inhospitable to most freshwater species.

Wetlands decrease even further in the Upper Estuary and are divided about equally between marsh and wet meadows. Conditions in the upstream part of the Upper Estuary are such that only a few animal species find it a suitable habitat. Free of competitors, some planktonic crustaceans (*Neomysis americana* and *Mysis stenolepis*) reach very high densities. Some fish, such as Atlantic tomcod and Rainbow smelt, spend an important phase of their life cycle in the salinity transition zone. Nonetheless, species diversity drops significantly. Less than a dozen freshwater fish species, and about the same number of saltwater species, inhabit this part of the St. Lawrence. Almost all benthic invertebrates have disappeared. Only aquatic shoreline plants show any diversity; about 80 species have been inventoried. This exceeds the number of species upriver (about 60) and downriver (about 45). However, many of these plants are at the limit of their salinity tolerance and their numbers are small. Plants best adapted to conditions in the Upper Estuary, such as bulrush and arrowhead, form extensive coastal marshes in the North Channel of Île d'Orléans, at Cap-Tourmente, Cap-Saint-Ignace, l'Islet-sur-Mer and in the Montmagny archipelago. These areas play a unique role in sediment dynamics and the annual cycle of the Greater snow goose.

Human population densities are low along the 150-km shoreline of the Upper Estuary and the Saguenay. One of the main commercial ports on the St. Lawrence is located in this stretch, at Baie-des-Ha! Ha! There is a strong tourism industry in the Charlevoix and Saguenay Fjord regions.

Two of the fifty industrial plants targeted under the St. Lawrence Action Plan operate in the main axis of the Upper Estuary, while eight others operate in the upstream part of the Saguenay. The plants are in the pulp and paper, metallurgy and mining sectors.

1.4 THE LOWER ESTUARY AND GULF

The St. Lawrence becomes much wider in the Lower Estuary and Gulf. This hydrographic region stretches over 230 km to Pointe-des-Monts and flows into the Gulf, a kind of semi-enclosed sea that contains several large islands: Anticosti, de la Madeleine, Bonaventure and Percé Rock, the Mingans and Sept-Îles.

About ten major tributaries flow into this section of the St. Lawrence, including the Manicouagan, aux Outardes, Moisie, Matane and Rimouski rivers. The circulation and mixing of waters in this marine environment is complex. In summer, the water is characterized by three superimposed layers of different temperatures and salinity levels; in winter there are two layers.

Wetlands here are composed mainly of wet meadows. Fish are once again abundant and extraordinarily diverse in form and adaptation: close to 100 species inhabit the Lower Estuary and the Gulf. The aquatic plant species along the shoreline are completely different from those upriver: only species that can bear the brunt of storms, ice and tides can survive. About 40 species manage to do so, the most abundant being *Spartina alterniflora* (cordgrass). Marine mammals appear and over a dozen species are commonly found. Oceanographic conditions in areas near the mouth of the Saguenay and the Mingan islands favour the formation of large swarms of planktonic shellfish and forage fish, creating an attractive food source for whales. Several species of marine birds become abundant toward the Lower Estuary, and Bonaventure Island in the Gulf has the largest colony of Northern gannets in North America.

This part of the St. Lawrence includes the commercial ports of Port-Cartier and Sept-Îles – which top the list of ports handling more than one million tonnes of goods per year – as well as the ports of Baie-Comeau, Havre-Saint-Pierre and Cap-aux-Meules.

The largest catches of the commercial marine fishery in the St. Lawrence are taken in this river sector, specifically in Gaspésie, Îles de la Madeleine and the Côte-Nord. The region also boasts a number of fishing harbours and sixty processing plants. Shellfish gathering is also becoming an increasingly popular activity for locals and tourists alike.

Riverside communities are few and there is little heavy industry, except for a pulp and paper mill and an aluminum smelter at Baie-Comeau, both among the 50 industrial plants targeted by the St. Lawrence Action Plan.

The features of the four hydrographic regions are summarized in Table 1.1. The maps are based on information in Appendix 1 – Fundamental Features of the St. Lawrence. Among other things, the map shows the boundaries of the main water masses and aspects of the biological communities, including species diversity of fish and vascular plant communities along the shoreline, regions with significant concentrations of marine mammals and aquatic birds, types of dominant littoral vegetation, and spawning grounds. It also provides important socio-economic data.

TABLE 1.1

Features of the hydrographic regions

FEATURES	HYDROGRAPHIC REGION			
	Fluvial Section	**Fluvial Estuary**	**Upper Estuary and Saguenay**	**Lower Estuary and Gulf**
River bed				
Main sedimentation zones	Lake Saint-François Lake Saint-Louis La Prairie basin Lake Saint-Pierre	Sheltered areas, bays, ports and marinas	Cap Tourmente flats Montmagny archipelago (strong turbidity between Île d'Orléans and Île aux Coudres)	Protected port basins Network of deepwater channels
Water				
	Distinct water masses	Distinct water masses until Portneuf	Freshwater/saltwater mixing zone	*In spring*: Three superimposed water masses *In winter*: Two distinct water masses
Number of tributaries	38	25	47	134
Mean annual discharge (m^3/s) **– Source:**	7800 at Cornwall – Great Lakes: 100%	11 500 at Trois-Rivières – Great Lakes: 68% – Tributaries: 32%	14 100 at La Malbaie – Great Lakes: 55% – Tributaries: 45%	16 800 at Baie-Comeau – Great Lakes: 46% – Tributaries: 54%
Salinity characteristics	Fresh water	Fresh water	Brackish and salt water	Salt water
Biological resources				
Wetland types	Aquatic grasses, marshes, wet meadows, swamps	Marshes, wet meadows, swamps	Marshes, wet meadows	Marshes, wet meadows
Biological diversity	Most diversified living communities; richest stretch: Lake Saint-Pierre sector.	Less biological diversity than is found in the Fluvial Section and Lower Estuary (including the Gulf), especially in terms of bacteria, vascular plants, zooplankton, benthic invertebrates and fish.		Increased biological diversity, particularly for fish, birds and marine mammals.
Uses				
Riverside population density in 1991 (inhab/km^2)	Most urbanized area: 749	213	38	3

TABLE 1.1 (cont'd)

Features of the hydrographic regions

FEATURES	HYDROGRAPHIC REGION			
	Fluvial Section	**Fluvial Estuary**	**Upper Estuary and Saguenay**	**Lower Estuary and Gulf**
Uses				
Commercial ports handling over 1 000 000 tonnes of goods in 1992 (rank out of 12)	Montreal-Contrecoeur (3) Sorel (6)	Quebec City (4) Bécancour (9) Trois-Rivières (10)	Baie-des-Ha! Ha! (7)	Port-Cartier (1) Sept-Îles (2) Baie-Comeau (5) Havre Saint-Pierre (8) Cap-aux-Meules (11)
Commercial fishing (based on 1991 landings)	Lake Saint-Pierre is the last bastion of commercial freshwater fishing	–	–	Area where commercial fishing is practised the most, particularly in Gaspésie, Îles-de-la-Madeleine and Côte-Nord.
Sport hunting and fishing	Freshwater sport fishing most intense here, especially on lakes Saint-Louis and Saint-Pierre. Highest waterfowl harvest.	–	–	–
Industrial activities	Most industrialized area			
– Number of designated priority plants under the St. Lawrence Action Plan	27	10	10	2
– Number of priority plants, by industrial sector	Inorganic chemicals (9) Organic chemicals (10) Metallurgy (6) Pulp and paper (2)	Inorganic chemicals (1) Organic chemicals (1) Metallurgy (2) Pulp and paper (6)	Inorganic chemicals (1) Organic chemicals (0) Metallurgy (3) Pulp and paper (6)	Inorganic chemicals (0) Organic chemicals (0) Metallurgy (1) Pulp and paper (1)
Utilization of banks*				
– Wetlands:	7.7%	4.0%	3.4%	N/A
Forest:	19.8%	43.7%	66.0%	
Farmland and idle land:	56.4%	41.4%	28.0%	
Built-up zone and exposed soil:	16.1%	10.9%	2.6%	

N/A: Not available.

**Only the mouth of the Saguenay River was considered for the Upper Estuary/Saguenay sector.*

2

Tracking the Health of the St. Lawrence

Influential Characteristics and Indicators of the State of the St. Lawrence

In order to produce an assessment of a system in which all elements constantly interact, researchers had to develop a simplified representation of the functioning of the St. Lawrence ecosystem. This is an essential step in the development of an accessible – yet non-reductive – synthesis of knowledge about the St. Lawrence.

Sector-by-sector studies on the physico-chemical, biological and socio-economic aspects of the St. Lawrence have few common denominators. The environment is complex, its components highly diverse, and the interactions among them poorly understood.

Through the active participation of some fifteen experts, a representation of how the St. Lawrence ecosystem works was developed in three stages using the Delphi method. This method is often used in situations where multiple variables and interrelationships require the synthesis of vast amounts of knowledge. The method involves having a group of specialists discuss opinions until they arrive at a consensus.

Fourteen characteristics were selected as being most significant for evaluating the condition of the St. Lawrence. Each characteristic describes aspects of the St. Lawrence that are important to better understanding its functioning, and identifies the elements that should be monitored to determine the health of the ecosystem, much the same way that organs and systems in the human body are analysed to diagnose the health of the body as a whole.

As in the human body, where all organs and systems are interrelated, the influence of each characteristic on the other characteristics was assessed qualitatively, in order to develop a basic picture of how the St. Lawrence ecosystem works. The characteristics were charted on a large matrix and their respective influences evaluated. With this model, it was possible to identify those characteristics of the ecosystem most influenced by the action of other characteristics, those that exert the most influence on the ecosystem, and those that both exert influence and are themselves influenced (hybrids), both for the St. Lawrence as a whole and for each hydrographic region.

The parallel with the human body goes even further, the difference being that while the human body often demonstrates obvious signs of illness or malfunction (pain, fever, anemia, tumours, etc.), the same is not true of a fluvial ecosystem. In addition, after the most significant characteristics were selected, they had to be "made to speak." This meant identifying a series of indicators which, when linked with various characteristics, could be used to assess the state of the environment.

This representation of these interrelationships, supplemented by indicators, became the basis for assessing the state of the St. Lawrence. Details on the methodology, criteria used to select characteristics, the representation of the functioning of the fluvial ecosystem, and a discussion of the indicators used to measure the state of the environment are presented in Part 4 of Volume 1 of this report.

The ecosystem concept

Before beginning a systematic review of the state of the St. Lawrence, terms and concepts should be clearly defined:

- All the living organisms and the physico-chemical environment in which they live, as well as their many interactions, constitute an ecosystem.
- In the *State of the Environment Report on the St. Lawrence River*, humans are considered an integral part of the ecosystem because the St. Lawrence plays a crucial role in the social, cultural and economic life of human communities along the River.
- Conversely, the quantity and quality of the water and biological resources of the River are largely dependent on the activities of riverside populations, including shipping, uptake of water, wastewater discharge, fishing, swimming and water sports.

2.1 CHARACTERISTICS SELECTED

Characteristics considered most indicative of the state of the St. Lawrence were selected using the following criteria:

1. The characteristic changes with time and space (it cannot be a constant).
2. The characteristic is found throughout the St. Lawrence.
3. The characteristic has a *direct* influence on the fluvial ecosystem.
4. The existence of historical data is an asset to the selection of a characteristic, but the absence of such data does not exclude a characteristic.

The characteristics were then grouped together according to four broad components of the St. Lawrence ecosystem:

1. River bed
2. Waters of the St. Lawrence
3. Biological resources
4. Uses associated with human activity.

Each component had a very high number of potential characteristics. When selection criteria were applied, that number was reduced to 14. Each of the 14 characteristics is considered of equal relative importance in terms of the mass of information available for the fluvial ecosystem as a whole. The characteristics were therefore not weighted in relation to each other.

The characteristics selected for each river component are identified in Table 2.1.

TABLE 2.1

The 4 components and 14 characteristics selected for assessing the state of the St. Lawrence

Components	**Characteristics**
River bed	1. Sediment quality
Waters of the St. Lawrence	2. St. Lawrence water quality
	3. Tributary water quality
Biological resources	4. Biodiversity
	5. Natural environments and protected species
	6. Condition of biological resources (abundance and contamination)
Uses associated with human activity	7. Shipping
	8. Modification of the floor and of hydrodynamics
	9. Shoreline modifications
	10. Urban wastewater discharges
	11. Industrial wastewater discharges
	12. Commercial fishing
	13. Sport hunting and fishing
	14. Accessibility of the banks and River

The 14 characteristics selected for assessing the state of the St. Lawrence were then placed in a grid called the "influence matrix." To complete the representation of the functioning of the St. Lawrence system, the intensity of the direct links between these characteristics was assessed qualitatively. See Chapter 3 for a description of each characteristic.

2.2 INTERACTIONS

If they could speak, each of the 14 characteristics would undoubtedly express concerns about being or not being influenced, and about having influence or not! Which indicates, of course, that the 14 characteristics are in constant interaction with each other.

Matrix analysis of the direct interactions between characteristics of the St. Lawrence ecosystem showed which characteristics are most heavily influenced and which are most influential. Some fall into both groups, being strongly influenced and also exerting considerable pressure on other characteristics.

The term "influence" was chosen because it seemed most appropriate for describing the observed links between characteristics. "Influence" depends on several factors; the effect on a characteristic may differ depending on the number, type and intensity of the pressures exerted upon it, and on its capacity for assimilation and recovery. Its condition may be either disturbed by the influence of another characteristic, or partially protected by it; or its equilibrium may be threatened by the intensity of one or more aggressive influences. Similarly, the influence one characteristic has on another is not always harmful, as for example with natural environments and protected species, which preserve biological resources and have a direct impact on biodiversity, access to the River and its banks, and sport hunting and fishing. In the case of biological resources, their condition could have a positive or negative influence on commercial fishing.

The characteristics selected and their influence on the St. Lawrence system as a whole are shown in Figure 2.1. "Influence" varies significantly in the various parts of the St. Lawrence. A characteristic that may be influential in the Fluvial Section might be completely dominated by other characteristics in the River as a whole. For example, sediment quality has a major influence on the condition of biological resources, but its influence diminishes considerably heading downriver from the Upper Estuary. Sport hunting and fishing, as well as biodiversity, are mainly influential in the main course of the River.

2.2.1 Influenced characteristics

Among the 14 characteristics selected for monitoring changes in the state of the St. Lawrence, the influenced characteristics are linked to the main findings and can also be a practical tool to measure the overall effectiveness of intervention

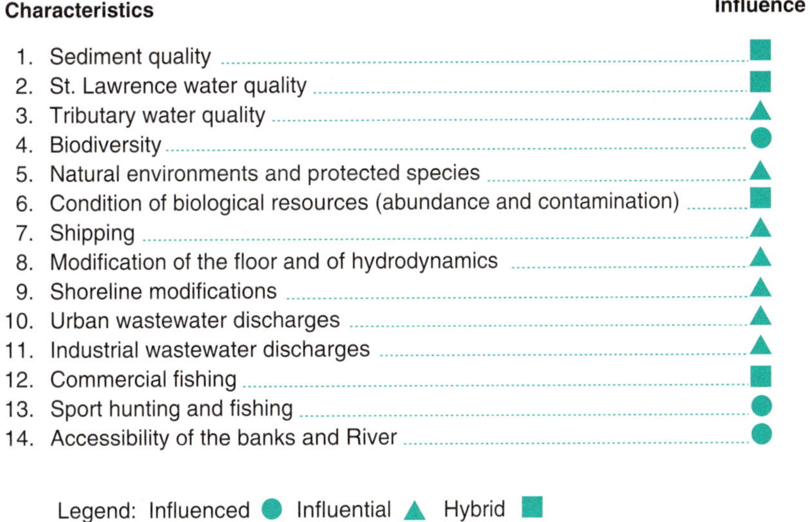

FIGURE 2.1

The 14 characteristics and their influence on the St. Lawrence as a whole

work, using measurements taken over a given period of time. They are listed in decreasing order of importance in Table 2.2.

These characteristics are subject to the most pronounced influences in the Fluvial Section and Fluvial Estuary. The further downstream one goes, the less these characteristics are affected by the number and intensity of influences, except for commercial fishing, which in every hydrographic region is strongly influenced by the condition of biological resources.

TABLE 2.2

The seven most-influenced characteristics for the St. Lawrence as a whole (in decreasing order of importance)

Characteristics	Importance
Condition of biological resources (abundance and contamination)	●
Accessibility of the banks and River	●
Sediment quality	●
St. Lawrence water quality	●
Sport hunting and fishing	●
Commercial fishing	●
Biodiversity	●

For the River as a whole, the influences that affect biodiversity are not evident. This is because not enough is known about biodiversity, and because biodiversity is closely associated with the condition of biological resources and shoreline modifications. Lastly, nothing is known about the strength of the influence exerted by the condition of biological resources on natural environments and protected species.

2.2.2 Influential characteristics

Because they can be acted on directly, characteristics that influence conditions help direct intervention work in the St. Lawrence. Eight characteristics are highly influential. They are listed in Table 2.3, in decreasing order of importance.

TABLE 2.3

The eight most influential characteristics for the St. Lawrence as a whole (in decreasing order of importance)

Characteristics	Importance
Industrial wastewater discharges	▲
Tributary water quality	▲
St. Lawrence water quality	▲
Natural environments and protected species	▲
Condition of biological resources (abundance and contamination)	▲
Urban wastewater discharges	▲
Commercial fishing	▲
Shoreline modifications	▲

Note: Shipping, sediment quality and modification of the floor and of hydrodynamics are not included among the characteristics with the most influence on the ecosystem as a whole because their main areas of influence are the Fluvial Section and Fluvial Estuary.

The network of relationships shows that interactions multiply in the two upstream sections and gradually decrease towards the Gulf. The Lower Estuary/Gulf sector has the fewest influences; furthermore, what influences there are have a much weaker impact, except for commercial fishing and condition of biological resources, which have a strong impact on each other.

2.2.3 Hybrid characteristics

Four of the most influenced characteristics for the St. Lawrence as whole also have an impact on all or part of the ecosystem. For purposes of analysis, these characteristics are referred to as hybrids. They are listed in decreasing order of importance in Table 2.4.

Because of their strong and combined interactions in the ecosystem, a general improvement in these characteristics will signal an improvement in the general state of the St. Lawrence. Conversely, a disturbance in all four

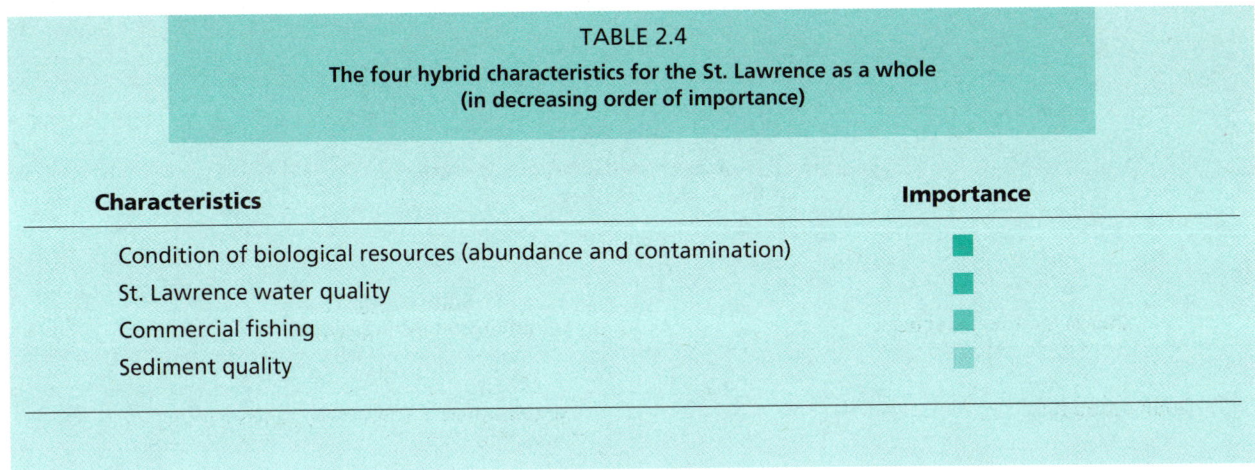

TABLE 2.4

The four hybrid characteristics for the St. Lawrence as a whole (in decreasing order of importance)

Characteristics	**Importance**
Condition of biological resources (abundance and contamination)	■
St. Lawrence water quality	■
Commercial fishing	■
Sediment quality	■

characteristics could signal a threat to ecosystem equilibrium. Losses related to use of river water would be cumulative, comprising loss of drinking water for direct human consumption, aquatic life, primary-contact recreation and commercial fishing, as well as loss of natural resources.

2.3 INDICATORS OF THE STATE OF THE ST. LAWRENCE

To assist in determining the condition of the 14 characteristics selected for assessing the state of the St. Lawrence and for direct environmental monitoring, a set of indicators was established on the basis of existing databases. The indicators are listed in Table 2.5.

Indicators selected for measuring the state of the St. Lawrence have limitations. It is not always possible to draw clear conclusions about the changing state of the environment or to encompass the St. Lawrence in its entirety. See Chapter 3 for a detailed discussion of the interpretation limits of the evaluation methods used and data available.

TABLE 2.5

State of the environment indicators for each of the 14 characteristics selected

Characteristics Selected	State of the Environment Indicators
1. Sediment quality	Percentage of exceedances of Toxic Effect Threshold (TET)
2. St. Lawrence water quality	Quality criteria exceedance index (CEI) for four uses
3. Tributary water quality	Chimiotox Index
4. Biodiversity	Number of wildlife and plant species at risk
	Introduction of exotic species (density of Zebra mussels on navigation buoys)
5. Natural environments and protected species	Surface area of protected habitats by category
6. Condition of biological resources (abundance and contamination)	Abundance of diverse species
	Contamination (flesh, liver or eggs) of diverse species (fish, birds, marine mammals)
7. Shipping	Total tonnage and percentage of dangerous goods
	Percentage of trips where vessel draft exceeds guaranteed minimum water depth
	Number of accidental spills reported
8. Modification of the floor and of hydrodynamics	Mean annual volume of dredged material
9. Shoreline modifications	Surface area of wetlands
10. Urban wastewater discharges	Percentage of riverside municipalities with wastewater treatment plants
	Percentage of population served by a treatment plant
11. Industrial wastewater discharges	Chimiotox Index
	Potential Ecotoxic Effects Probe (PEEP)
12. Commercial fishing	Landings (fresh and salt water)
13. Sport hunting and fishing	Harvest
	Catch (fish)
	Restrictions on consumption of fish
14. Accessibility of the banks and River	Number of recreation and tourism infrastructures
	Number of public beaches open

3

The Health of the St. Lawrence

Assessing the St. Lawrence River

The condition of the St. Lawrence was first assessed in 1977. Some 18 years later, it is interesting to look at the conclusions of the Comité d'Étude sur le Fleuve Saint-Laurent (St. Lawrence River study committee), a federal-provincial committee mandated to answer fundamental questions raised by management, development and protection of the St. Lawrence environment.

The conclusions of the *Interim Report on the St. Lawrence* filed by the Committee are rather sombre. Six categories of deterioration are reported:

- Dissemination of toxic substances
- Bacterial contamination
- Encroachment on wildlife habitats
- Destruction of aesthetic features
- Excessive suspended matter
- Enrichment by nutrients.

As perceived in the late 1970s by the Committee, the three main causes for the deteriorating condition of the St. Lawrence were the degradation of many of its tributaries, wastewater discharges and filling of shoreline.

Needless to say, the approach taken by the Committee in 1977 differed from the one applied in this report to assess the condition of the St. Lawrence ecosystem, but the information collected during the earlier period remains relevant nonetheless. Whenever data was available – which was not always the case – it served as a reference point for tracking changes in the condition of the River since the late 1970s.

Similarities and differences between the present day and the late 1970s appear throughout the analyses of the characteristics selected for this assessment of the River.

Contamination of wildlife and of sediment, exceedances of criteria for diverse water uses and restrictions on consumption of fish are some of the signs of disturbance of the St. Lawrence. The main findings of the analyses of the seven characteristics considered most affected show stable situations or changes that can mean increase or loss of resources and uses – which can be either positive or negative for the St. Lawrence environment. Thus, the summary presented in Appendix 2 shows that:

- Abundance has been maintained for the marine mammal considered in the estuary and Gulf;
- Contamination declined for the two species considered in the Lower Estuary and Gulf;
- Public beaches along the St. Lawrence River are open; and
- The coastal areas of Bas-Saint-Laurent and the Gaspé are problematic for shellfish-harvesting.

These findings are thus matters for concern when it comes to the state of health of the St. Lawrence. Caution is required when interpreting the findings, however. Increase in a population (Greater snow goose, for example) does not necessarily mean the population is doing better or inhabiting a healthy ecosystem; the increase may be a reaction to a problem in the ecosystem.

The main causes of changes in the condition of the St. Lawrence are:

- Large quantities of toxic substances
- Omnipresent pathogenic bacteria (fecal coliforms)
- Harvesting pressure on certain fish species in difficulty or in serious decline
- Inadequacy of protected natural environments to offset physical modification of habitats.

The information collected on the characteristics identified as having the greatest influence on the St. Lawrence ecosystem suggest five of the characteristics are mainly responsible for the changes: industrial wastewater discharges, tributary water quality, natural environments and protected species, urban wastewater discharges, and commercial fishing. The map in Appendix 3 shows the main pressures on the St. Lawrence.

The pages that follow detail results of analyses of the 14 characteristics using different indicators.

The process is the same for each characteristic:

1. Identification of the characteristic;
2. Identification of indicators used;

3. Background information, including the reasons for selecting the characteristic and the indicators;
4. General interpretation of the state of the characteristic in light of all indicators used to monitor it;
5. Analysis of indicators used – definitions, limitations, results (with tables and figures when appropriate) – and pinpointing of problem areas or areas at risk;
6. Information supplements, when appropriate;
7. Wrap up: Main findings, relative importance of each characteristic compared to the other characteristics and recommendations to improve monitoring.

READER'S GUIDE

Refers the reader to appendix map 2 or 3.

Signifies the presentation of an indicator associated with a characteristic.

Refers the reader to a given section (2.4.2.1) of Volume 1 (V1) in parts 1: Physico-Chemical Aspects (P1), Part 2: Biological Aspects (P2), or Part 3: Socio-Economic Aspects (P3).

● Influenced
▲ Influential
■ Hybrid

} Refers to the nature of the characteristic (the darker the symbol, the greater its relative importance).

3.1

SEDIMENT QUALITY

Danielle Gingras

Background

Most contaminants have a chemical affinity for settled particulate matter (sediments), with the result that they may concentrate 100 000 times more toxic substances than the surrounding water. Certain substances (pesticides, organochlorines, PAHs, PCBs and heavy metals) persist long after the source of contamination has been eliminated, and some modes of sediment transport cause contaminants to return to the water column and to enter the food chain.

Interim criteria for assessing the quality of surface sediment were developed by Environment Canada and the Ministère de l'Environnement du Québec (MENVIQ). These criteria define three levels of effect that contaminants can have on benthic organisms: No Effect Threshold, Minimal Effect Threshold and Toxic Effect Threshold. The percentage of exceedances of the Toxic Effect Threshold (TET) was selected as an indicator for assessment purposes.

Interpretation

Trends in sediment contamination in the Fluvial Section and the Fluvial Estuary can be analysed by comparing 1976 data with data from 1984 to 1992. Historical analysis of sediment contamination by the seven metals analysed generally shows a substantial decrease in heavy-metal contamination of surface sediment in the Fluvial Section between 1976 and 1992, and low contamination in the Fluvial Estuary. Chromium, mercury, copper and nickel contamination are still, nonetheless, problematic in the main course of the River. In addition, areas of contamination differ greatly from one contaminant to the next depending on the year and sources of contamination.

TET exceedance frequency

Definition

The Toxic Effect Threshold (TET) of a particular contaminant is the critical concentration at which most benthic organisms (90%) suffer major damage. TET exceedance frequency is the ratio of values exceeding TET to the total number of analyses expressed as a percentage, and it is used to identify sedimentation areas that are most contaminated and most affected by toxic substances. Variation in exceedance frequency over a period of time indicates change in surface sediment quality. Available data cover the period from 1976 to 1992.

Limitations

At the moment, this indicator offers only a partial picture of the situation in the St. Lawrence because it is affected by sampling effort, which has focused on the fluvial lakes, and because it only takes the following seven metals into account: cadmium, copper, chromium, mercury, nickel, lead and zinc. For the Upper Estuary, Lower Estuary and Gulf, in particular, it was difficult to arrive at overall assessments: for these parts of the St. Lawrence, data obtained over a five-year period had to be grouped together to obtain minimum spatial coverage. Moreover, no adjustments were made for particle-size composition of the sediment, and the total form of the metal was investigated.

Problem Or Risk-Prone Sectors

*Lake Saint-Louis, lesser La Prairie basin, area upstream of Lake Saint-Pierre and the ports of Montreal and Quebec City.

**The area between Quebec City and Île aux Oies.

Results

Since 1976, 5963 analyses of sediment quality in the St. Lawrence have been conducted, 2618 in 1976 (the Fluvial Section and the Fluvial Estuary only) and 3345 between 1984 and 1992 (all sectors of the St. Lawrence). The majority of TET exceedances since 1976 were recorded in the Fluvial Section (559/692, or 81% of exceedances recorded). Seventy-five percent of TET exceedances between 1984 and 1992 (161/215) were recorded in the Fluvial Section or the Fluvial Estuary (see Figure 3.1.1). Compared to the St. Lawrence as a whole, exceedance frequencies are low in the Upper Estuary, Lower Estuary and Gulf. However, in these three sectors of the River, sediment quality was analysed almost exclusively around port facilities, and hence the analyses do not indicate sediment quality throughout the environment; at most they indicate their degree of contamination in areas of industrial and port activity.

Contamination of the most upstream reach of the Fluvial Section (Lake Saint-François) dropped dramatically between 1976 and 1992 for all metals. In fact, no exceedances of TET have been recorded in Lake Saint-François since

FIGURE 3.1.1

Toxic Effect Threshold (TET) exceedance frequencies (%) for seven contaminants, by river sector (1976-1992)

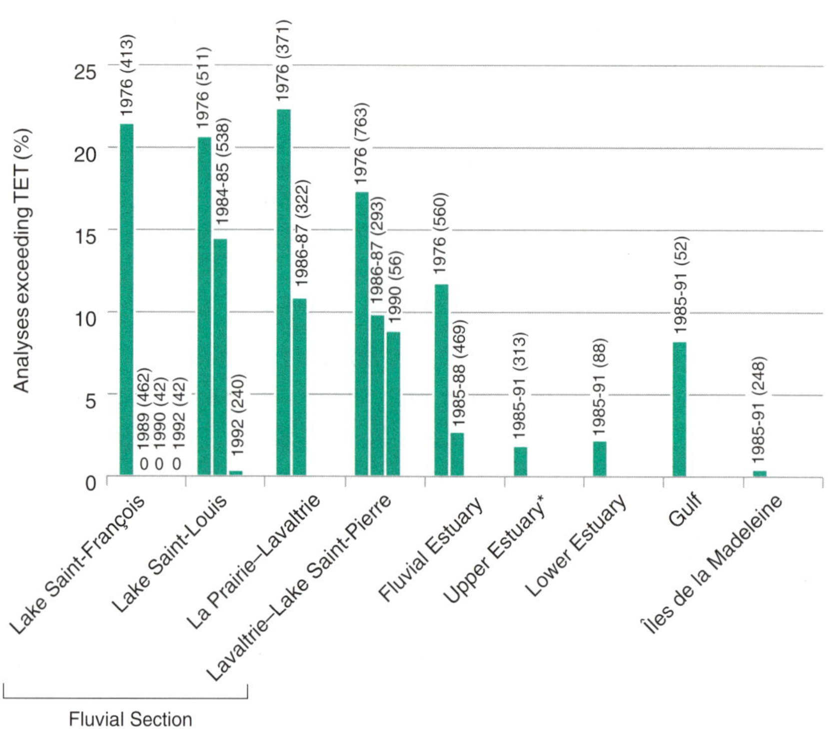

Note: *Excluding the Saguenay River.

Contaminants measured: Cd, Cr, Cu, Hg, Ni, Pb, Zn.

Figures above bars indicate sampling year; those in brackets indicate number of analyses.

Source: Based on data from Olivier and Bérubé, 1993; Lorrain and Pelletier, 1993.

1989. The decrease in contamination was due to a major cutback in discharges to the Cornwall area with the enforcement of regulations governing industrial effluents as well as measures taken in the Massena area to reduce toxic inputs.

In Lake Saint-Louis, sediment quality generally improved between 1976 and 1985, especially for cadmium, mercury and lead contamination. Chromium contamination, on the other hand, increased substantially in Lake Saint-Louis. And despite the drop in mercury concentration, the lake was still highly contaminated by mercury in 1984-1985. Sediment quality continued to improve there between 1985 and 1992; by 1992, only one analysis out of 240 exceeded TET.

In the La Prairie–Lavaltrie area, sediment contamination by chromium and zinc increased appreciably, whereas contamination by cadmium, mercury, copper, nickel and lead diminished.

In the Lavaltrie–Lake Saint-Pierre outlet area, sediment contamination by chromium and copper increased whereas contamination by cadmium, mercury and nickel diminished.

The Fluvial Estuary is less contaminated than the Fluvial Section. There are nonetheless problems with chromium, copper, mercury and nickel contamination, mainly at the mouth of the Saint-Charles River, at Quebec City, though data for this date back several years.

Sediment quality in the Upper Estuary, Lower Estuary and Gulf is better than sediment quality in fresh water. The Upper Estuary, however, does act as a sink: most of the trace metals come from the Fluvial Estuary, are adsorbed onto fine suspended matter in the maximum turbidity zone and settle on the river bottom due to ambient conditions favouring sedimentation. Few trace metals thus make it to the Gulf. Sediment quality in salt water ranges from very good to passable, except close to the shores of certain municipalities and around certain port facilities, where marked sediment enrichment has been observed. Sandy Beach harbour – where large quantities of copper ore have been dumped, contaminating harbour sediment – is a case in point.

SEDIMENT QUALITY

Main Findings

- Since 1976, surface sediment sampling efforts have focused heavily on the fluvial lakes, making it difficult to get a picture of the St. Lawrence as a whole. A dramatic drop in surface sediment contamination by the heavy metals investigated was nonetheless reported in the Fluvial Section.
- Surface sediments in certain stretches of the Fluvial Section and the Fluvial Estuary are still highly contaminated: Lake Saint-Louis, the lesser La Prairie basin, the upstream part of Lake Saint-Pierre, the area between Quebec City and Île aux Oies, and the ports of Montreal and Quebec City.

Relative Importance

- Sediment quality ranks last of the hybrid characteristics. It has a considerable impact on bioresources, but the impact decreases seaward starting at the Upper Estuary.

Recommendations on Monitoring

- Improvement of the TET exceedance indicator to take into account sediment particle-size composition should be considered, or perhaps development of a new indicator to make it easier to obtain a picture of the St. Lawrence as a whole.
- Periodic measurements that take sediment dynamics into account are recommended, as is data collection in areas for which there is very little data. For the moment, there are no indicators which incorporate amplitude and frequency of exceedance that can be used to obtain overall assessments of sediment quality in each of the sectors of the St. Lawrence.

3.2

ST. LAWRENCE WATER QUALITY

St. Lawrence Centre, C. Hudon

Background

A detailed diagnosis of water quality must consider two main factors: the transition from fresh to salt water upstream to downstream in the St. Lawrence River, and the diversity of water uses. Based on available information, eight indicators were selected to assess freshwater and saltwater quality for four specific water uses.

The freshwater uses are as follows:

- Aquatic life (two indicators)
- Direct human consumption (three indicators)
- Primary-contact recreation (two indicators).

The saltwater use is as follows:

- Shellfish harvesting (one indicator).

Interpretation

Aquatic life

Low exceedances of both organics and inorganics aquatic life criteria were observed at the entrance to the St. Lawrence, in the waters of the Great Lakes and in the waters of the Ottawa River. Water quality for aquatic life deteriorates downstream of Montreal (especially in the mixed waters along the north shore) and then improves slightly in the mixed waters of the Quebec City region. This trend seems consistent with the increase in turbidity and suspended solids from upstream to downstream and the presence of soft water masses. The five variables that pose a potential risk to aquatic life are copper, chromium, lead, DDT and PCBs.

Direct human consumption

Fresh water from the St. Lawrence cannot be consumed raw because of its poor bacterial quality, due mainly to the presence of fecal coliforms. No restrictions are imposed because of inorganics content, however. Data on the quality of water for direct human consumption – that is, untreated water – show exceedances of bacterial quality criteria throughout the St. Lawrence in 1991. On the other hand, none of the organic variables investigated exceeded direct human consumption water quality criteria in 1991. As for inorganic variables, exceedances of iron, turbidity and manganese criteria were recorded between 1985 and 1990. These variables must be reduced for aesthetic reasons that can interfere with human consumption of water.

Primary-contact recreation

The highest counts of coliforms in recreational waters were registered in the main course of the River. Data from 1990 to 1993 show a major problem, mainly east of Montreal Island to Quebec City. High fecal coliform counts increase the probability of finding pathogenic elements (viruses and bacteria) in the water that directly affect human health.

Shellfish harvesting

In 1992, bacteriological standards were most often exceeded at stations in the Gaspé and the Bas-Saint-Laurent. Because of bacterial contamination, among other things, 40 and 26 shellfish areas, respectively, were closed in these two regions. For lack of information, the quality of the marine environment for other uses such as aquatic life and primary-contact recreation as a function of the presence of inorganics or organics cannot be evaluated.

Fresh Water

Aquatic life

Organics criteria exceedance index (OCEI) and inorganics criteria exceedance index (ICEI)

Definition

The quality criteria exceedance index (CEI) takes into account mean amplitude of quality criteria exceedance of each variable, frequency of criteria exceedance (expressed as a percentage) and the number of variables considered. The higher the CEI, the poorer the quality of the water for a given use.

Two CEIs were calculated for aquatic life, one for organics (OCEI) and one for inorganics (ICEI).

These indexes of exceedance of chronic toxicity criteria indicate the capacity of a waterway to welcome and sustain aquatic life. The indexes are calculated using 10 organics (for the OCEI) and 12 inorganics (ICEI).

Limitations

These indexes give only a partial picture because chronic toxicity criteria for aquatic life only take into consideration parameters (a certain number of substances) that can be measured with conventional analytic methods (variable detection thresholds). The impact of chronic toxicity due to trace contaminants on aquatic organisms is still poorly understood, and the criteria do not take synergistic or antagonistic effects of the mixing of diverse substances into account.

PROBLEM OR RISK-PRONE SECTORS

Results

Figures 3.2.1 and 3.2.2 show OCEIs calculated based on data from 1991 and ICEIs for data from 1985 to 1990 in the water masses between Cornwall and Quebec City. Upstream of Montreal, the water is only slightly altered, but it deteriorates downstream – especially along the north shore – and then improves slightly in the mixed waters of the Quebec City region.

OCEIs are very low in the freshwater parts of the St. Lawrence, 0.00 to 0.41. OCEIs are very low upstream of Montreal but increase unequally in the water masses downstream of Montreal. The highest OCEI is in the Bécancour

FIGURE 3.2.1

Aquatic life: Organics criteria exceedance indexes (OCEIs) in the water masses between Cornwall and Quebec City (1991)

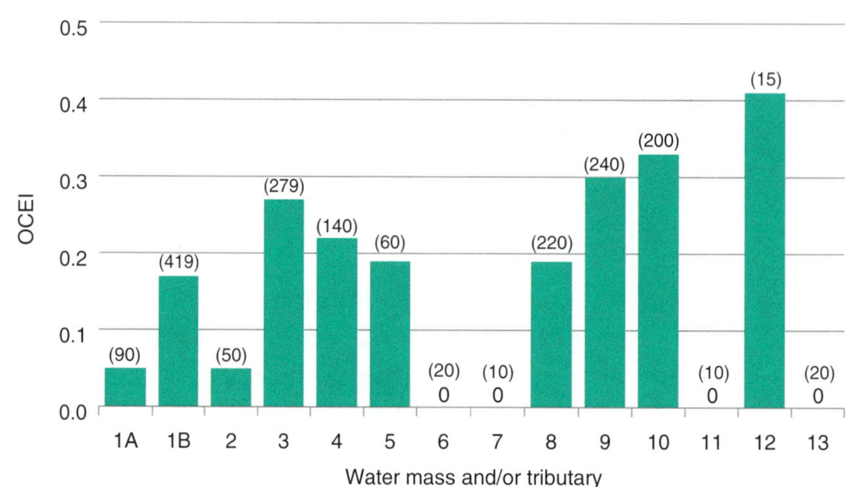

1A	Great Lakes–upstream of Montreal	7	L'Assomption River
1B	Great Lakes–downstream of Montreal	8	Richelieu River
2	Ottawa River	9	Yamaska River
3	Ottawa River–north shore mix	10	Nicolet River
4	Great Lakes–south shore mix	11	Saint-Maurice River
5	Quebec City region	12	Bécancour River
6	Saint-Louis River	13	Jacques-Cartier River

Note: Figures in brackets indicate number of samples.

Organics measured: endosulfan, PCBs, chlordane, total DDT, dieldrin, endrin, heptachlor and heptachlor epoxide, mirex, acenaphthene and naphthalene.

Source: Based on data from SLC, 1993a.

River water mass (0.41), followed by the Nicolet River water mass (0.33), the Yamaska River water mass (0.30) and the water mass of the Ottawa River–north shore mix (0.27). DDT and its metabolites and PCBs are the variables responsible for the exceedances. No exceedances of organics criteria (OCEI = 0) were recorded in the water masses of the Saint-Louis, L'Assomption, Saint-Maurice and Jacques-Cartier rivers.

As the figures show, ICEIs range from 0.11 to 1.33; in other words, a number of areas are highly degraded. The highest ICEI (1.33) was recorded in the water mass of the Saint-Louis River, followed by the water mass of the Ottawa River–north shore mix downstream of Montreal (after the confluence of Rivière des Prairies, Rivière des Mille Îles, L'Assomption River and the St. Lawrence), where copper and lead criteria were greatly exceeded and an ICEI of 1.24 was recorded. The Quebec City region comes next with an ICEI of 0.76. The Sorel area is also degraded: ICEI of the Great Lakes–south shore mix is 0.68 here. Exceedances of copper and chromium criteria are caused by industrial discharges in the region, the source of the degradation. The Great Lakes water mass upstream of Montreal is the water mass that least meets inorganics water quality criteria (ICEI = 0.11).

FIGURE 3.2.2

Aquatic life: Inorganics criteria exceedance indexes (ICEIs) in the water masses between Cornwall and Quebec City (1985-1990)

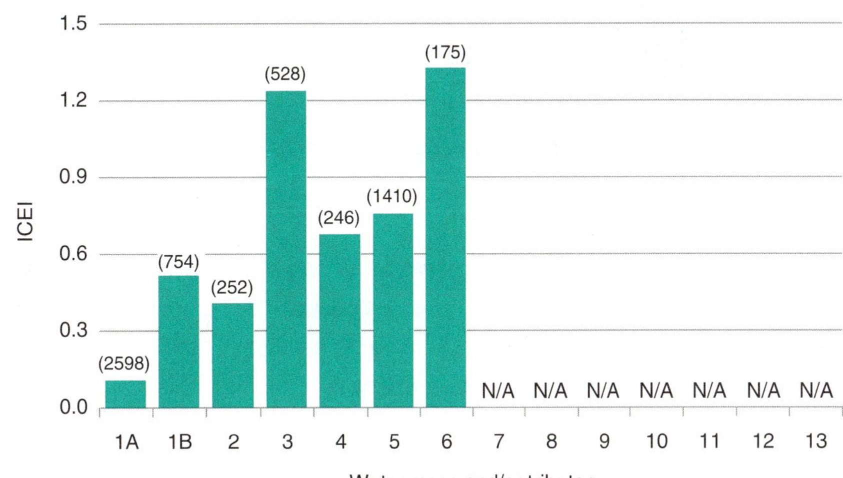

1A	Great Lakes–upstream of Montreal	7	L'Assomption River
1B	Great Lakes–downstream of Montreal	8	Richelieu River
2	Ottawa River	9	Yamaska River
3	Ottawa River–north shore mix	10	Nicolet River
4	Great Lakes–south shore mix	11	Saint-Maurice River
5	Quebec City region	12	Bécancour River
6	Saint-Louis River	13	Jacques-Cartier River

Note: Figures in brackets indicate number of samples.

N/A: Not available.

Inorganics measured: iron, nickel, copper, zinc, lead, aluminum, barium, cadmium, cobalt, chromium, molybdenum and vanadium.

Source: Based on data from SLC, 1993a.

Three inorganics are responsible for the exceedances that are potentially damaging to aquatic life: chromium, copper and lead. Criteria for certain metals change as a function of water hardness, however, which can affect metal bioavailability. In other words, copper and lead are likely to be more toxic in the brown waters along the north shore of the St. Lawrence, which are relatively soft, than in the other water masses. Other metals – iron and aluminum, for example – are also responsible for the fluctuations in ICEI, but since they are not very bioavailable and are strongly associated with particulate matter, they pose no chronic toxicity risk. In addition, under the pH conditions in the St. Lawrence (6.5 to 9.0), aluminum should not have any toxic effects.

Fresh Water

Direct human consumption

Inorganics criteria exceedance index (ICEI), organics criteria exceedance index (OCEI) and fecal coliform exceedance frequency

Definition

Two CEIs were calculated for direct human consumption, one for organics (OCEI) and one for inorganics (ICEI). Each of these criteria exceedance indexes for direct human consumption incorporates a number of variables (6 in the case of the OCEI and 12 in the case of the ICEI) associated with taste (organoleptic effects), appearance (aesthetic effects) and health (toxic effects).

Another index assesses bacterial quality only (fecal coliforms) by calculating direct human consumption water quality criterion (0 f.c./100 mL) exceedance frequencies. Measurements of bacterial quality indicate the intensity of treatment required to obtain water that is free of fecal coliforms and suitable for drinking.

Fecal coliforms serve as indicators of unhealthy conditions in the aquatic environment because they are associated with intestinal waste from warm-blooded vertebrates such as human beings. Fecal coliforms are thus indicators of the bacterial contamination of water. In aquatic environments, the most frequently encountered fecal coliforms are *Escherichia*, *Enterobacter* and *Klebsiella*. *Escherichia coli* alone accounts for 75% to 95% of fecal coliforms measured in the water. When *Escherichia coli* counts are too high, the probability of detecting pathogenic elements in the water (viruses and bacteria, for example) that directly affect primary-contact recreation and human health increases.

Limitations

Water quality criteria exceedance indexes for a given use and a given group of substances can be compared, but indexes for different uses cannot be compared, mainly because the number of parameters used is never the same.

The value of information obtained by measuring concentrations of toxic substances in water must not be overestimated. Such information must be supported by studies of their impacts on the environment – especially on users of the environment; that is, on human beings and aquatic organisms.

Despite their toxic potential, some substances – cyanides, volatile substances and phenols, for example – are not examined here because there was little data on their concentrations in the St. Lawrence.

Results

Problem Or Risk-Prone Sectors

OCEIs for drinking water for direct human consumption were calculated for 1991 only. Six organic variables were used: chlordane, total DDT, heptachlor and heptachlor epoxide, aldrin, dieldrin, and benzo(a)pyrene. Apart from one exceedance of benzo(a)pyrene at the mouth of the Saint-Louis River, all values obtained met the established criteria.

Figures 3.2.3 and 3.2.4 show ICEIs based on measurements taken between 1985 and 1990 in the water masses between Cornwall and Quebec City, as well as exceedances of the fecal coliform criterion in these water masses between 1990 and 1993.

As the figures show, ICEIs are low in the freshwater reaches of the St. Lawrence, ranging from 0.01 to 0.91. The water mass of the Saint-Louis River has the highest ICEI (0.91), followed by that of the Ottawa River–north shore mix (0.32). The Great Lakes water mass upstream of Montreal is the water mass that best meets inorganics water quality criteria (ICEI = 0.01).

FIGURE 3.2.3

Water for direct human consumption: Inorganics criteria exceedance indexes (ICEIs) in the water masses between Cornwall and Quebec City (1985-1990)

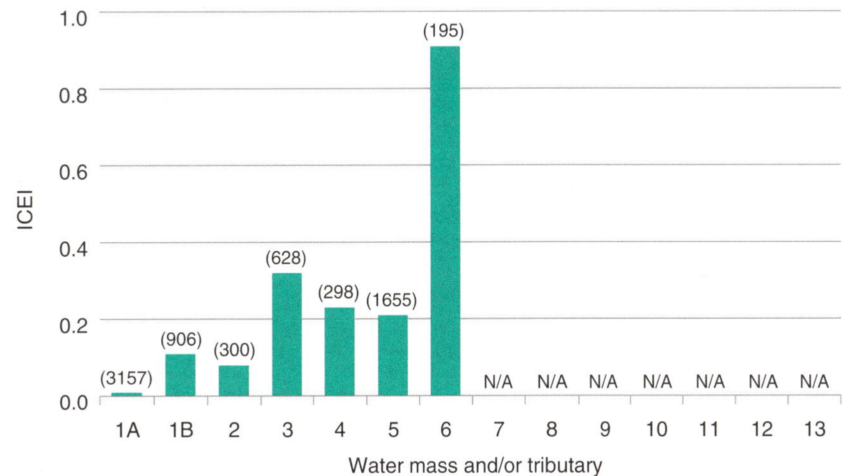

1A	Great Lakes–upstream of Montreal	7	L'Assomption River
1B	Great Lakes–downstream of Montreal	8	Richelieu River
2	Ottawa River	9	Yamaska River
3	Ottawa River–north shore mix	10	Nicolet River
4	Great Lakes–south shore mix	11	Saint-Maurice River
5	Quebec City region	12	Bécancour River
6	Saint-Louis River	13	Jacques-Cartier River

Note: Parameters measured: sodium, sulphates, turbidity, manganese, iron, copper, zinc, lead, barium, cadmium, chromium and chlorides.

Figures in brackets indicate number of samples.

N/A: Not available.

Source: Based on data from SLC, 1993a.

FIGURE 3.2.4

Water for direct human consumption: Fecal coliform criterion exceedance frequencies in the water masses between Cornwall and Quebec City (1990-1993)

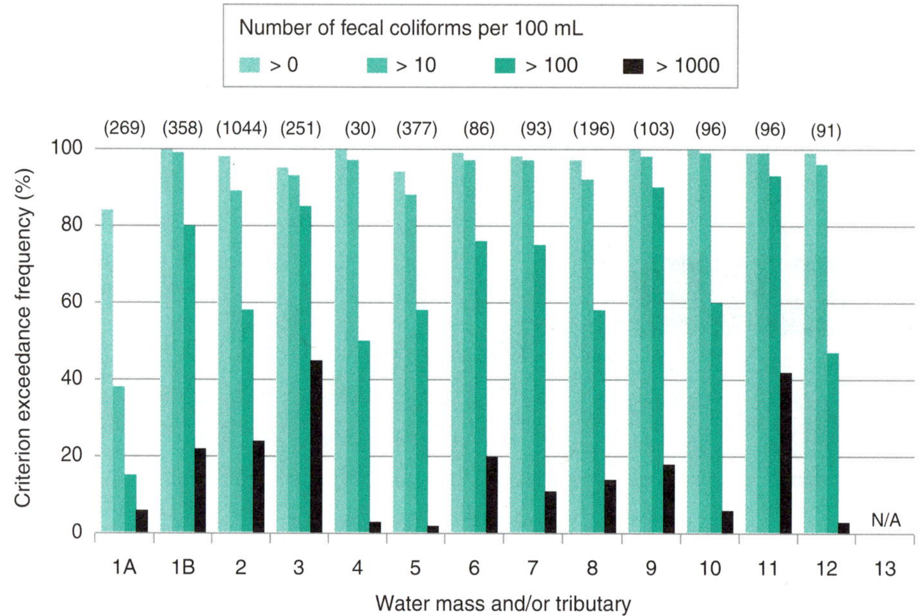

1A	Great Lakes–upstream of Montreal	7	Richelieu River
1B	Great Lakes–downstream of Montreal	8	Nicolet River
2	Ottawa River	9	Saint-Maurice River
3	Ottawa River–north shore mix	10	Bécancour River
4	Great Lakes–south shore mix	11	L'Assomption River
5	Quebec City region	12	Jacques-Cartier River
6	Yamaska River	13	Saint-Louis River

Note: Figures in brackets indicate number of samples.

N/A: Not available.

Source: Based on data from MENVIQ, 1993.

Untreated water from the St. Lawrence does not therefore contain inorganics in concentrations that render it unfit for consumption. The impact of the only exceedances noted – turbidity, iron and manganese – is aesthetic, not toxic. High concentrations of iron and manganese will stain clothing, damage plumbing and affect the taste of water. The highest ICEIs are at the mouth of the Saint-Louis River and at one station on the north shore, under Laviolette bridge in Trois-Rivières, because turbidity and iron content are high at these spots. Concentrations of other variables are well below quality criteria for direct human consumption.

Fecal coliforms were detected in almost all water samples analysed. Frequency of exceedance of drinking water criterion (0 f.c./100 mL) was more than 84% in all water masses. The water in the St. Lawrence cannot, therefore, be consumed directly – that is, untreated – because of its poor bacterial quality. The most problematic areas are the water masses of the Great Lakes–south shore mix, the Saint-Maurice and Bécancour rivers (where exceedance frequency is 100%) and the Great Lakes water mass downstream of Montreal (where

exceedance frequency is 99.8%). Very high fecal coliform counts (more than 1000 f.c./100 mL) are most frequent in the water masses of the Ottawa River–north shore mix and L'Assomption River.

Fresh Water

Primary-contact recreation

Inorganics criteria exceedance index (ICEI) and fecal coliform criterion exceedance frequency

Definition

Primary water contact recreation includes swimming and windsurfing.

One of the indicators used is the ICEI. This index includes a nutrient (total phosphorus), a physical parameter (turbidity) and eight inorganic substances. The indicator was calculated for the period 1985-1990.

The other indicator selected for assessment of recreational waters is frequency of exceedance of the fecal coliform criterion (200 f.c./100 mL) considered safe for the practice of primary-contact recreational activities. This indicator was calculated from data collected between May and October, 1990 to 1993.

Limitations

If water quality trends are to be identified, data must be adequately distributed over space and time, since it is difficult to interpret sharp fluctuations in bacteriological results over a short period. The data used, however, came from only a small number of stations in each water mass.

PROBLEM OR RISK–PRONE SECTORS

Results

Figure 3.2.5 shows ICEIs based on data from 1985 to 1990 in the water masses between Cornwall and Quebec City, as well as fecal coliform criterion exceedance frequencies between 1990 and 1993 in these same water masses.

As the figure shows, ICEIs for recreational waters are rather low, ranging from 0.07 to 0.67. The highest ICEIs were recorded in the water masses of the Saint-Louis River (0.67) and the Ottawa River–north shore mix (0.23). The water masses of the Great Lakes downstream of Montreal and of the Ottawa River are

FIGURE 3.2.5

Primary-contact recreational water: Inorganics criteria exceedance indexes (ICEIs) (1985-1990) and fecal coliform criterion exceedance frequencies in the water masses between Cornwall and Quebec City (1990-1993)

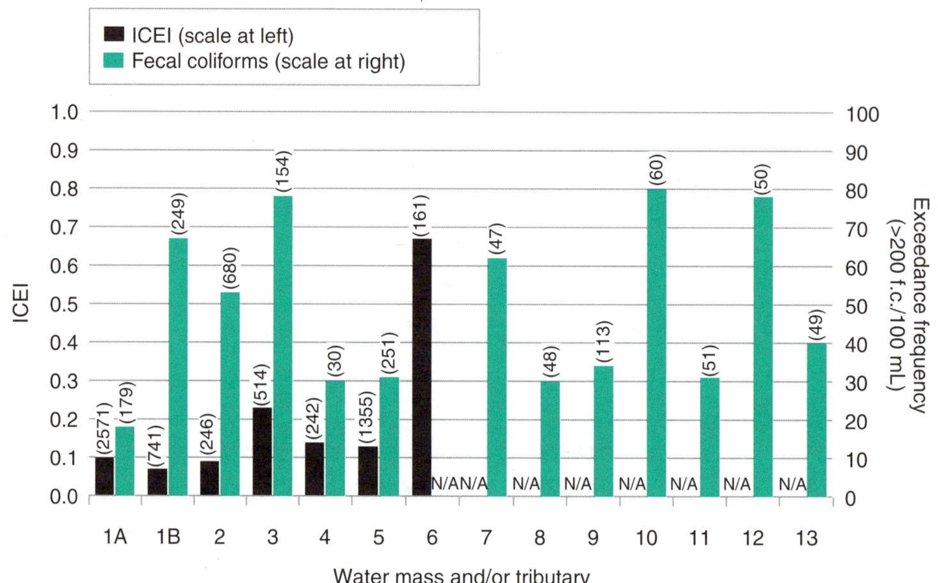

1A	Great Lakes–upstream of Montreal	7	Yamaska River
1B	Great Lakes–downstream of Montreal	8	Richelieu River
2	Ottawa River	9	Nicolet River
3	Ottawa River–north shore mix	10	Saint-Maurice River
4	Great Lakes–south shore mix	11	Bécancour River
5	Quebec City region	12	L'Assomption River
6	Saint-Louis River	13	Jacques-Cartier River

Note: Figures in brackets indicate number of samples.

N/A: Not available.

Inorganic variables measured: turbidity, phosphorus, barium, cadmium, chromium, copper, manganese, nickel, lead and zinc.

Source: Based on data from SLC, 1993a; MENVIQ, 1993.

those that best meet inorganics quality criteria (ICEIs are 0.07 and 0.09, respectively).

No exceedances of criteria were recorded in the waters of the St. Lawrence for the eight metals investigated. Metal concentrations do not therefore constitute an obstacle to primary water contact recreational activities. The parameters with the greatest impact on primary-contact recreational water quality in the St. Lawrence are total phosphorus and turbidity, which pose no health risks but can limit recreation for aesthetic reasons. Frequency of exceedance of the total phosphorus criterion (0.03 mg/L) is 30% in the water mass of the Great Lakes and considerably higher in the other water masses. Total phosphorus is not considered toxic in the strict meaning of the term, however. Between Cornwall and Quebec City, total phosphorus content increases from about 0.010 mg/L to 0.040 mg/L. This marked increase is due mainly to municipal wastewater discharges and leaching of phosphorus (fertilizers and manure) from farmland. The waters of the St. Lawrence are only slightly turbid upstream of Montreal, whether in the Great Lakes water mass or the Ottawa River water mass. Downstream of Montreal, turbidity increases along both the north and south shores, and particularly in the Lake Saint-Pierre region.

Fecal coliforms were detected in almost all water samples analysed. Frequency of exceedance of the recreational water fecal coliform criterion (200 f.c./100 mL) in the water masses studied ranged from 18% to 80%. Exceedance frequency was more than 60% in five water masses: those of the Saint-Maurice River (80%), L'Assomption River (78%), the Ottawa River–north shore mix (78%), the Great Lakes downstream of Montreal (67%) and the Yamaska River (62%). The most problematic areas are thus between the east end of Montreal Island and Quebec City.

Results for the Great Lakes water mass upstream of Montreal are affected by discharges of wastewater from southeast Montreal Island, which is not conveyed to a wastewater treatment facility. The substantial exceedances of fecal coliform criterion in the water mass of the Ottawa River–north shore mix are caused by the poor bacterial quality of the water from Rivière des Prairies and Rivière des Mille Îles and by wastewater treated by the Montreal Urban Community (MUC) and discharged to the St. Lawrence without disinfection. This influence extends right to the islands of Berthier–Sorel. In the water mass that flows along the north shore, the worst results were found at Trois-Rivières and immediately downstream in the water mass of the Saint-Maurice River. The situation improves in the mixed waters of the Quebec City region, which begin at Portneuf. At the mouths of the Richelieu, Nicolet, Bécancour and Jacques-Cartier rivers, recreational water quality is better than in the Quebec City region, where exceedance frequencies range from 30% to 40%.

Salt Water

Shellfish harvesting

Exceedance of fecal coliform standards

Definition

When the water of a shell bed is contaminated by fecal coliform bacteria, the shellfish living there absorb and concentrate the bacteria. The concentration of bacteria does not affect the shellfish, but it does make them unfit for human consumption.

Bacteriological analyses can give an assessment of salt water bacteria count. Results are expressed as the most probable number (MPN = number of coliforms per 100 mL) and are based on probability calculations. For an area to be open for shellfish harvesting, it must, among other things, meet the following standards: median coliform count at any station must not exceed 14 fecal coliforms per 100 mL of water; and no more than 10% of the counts at any station may exceed 43 fecal coliforms per 100 mL of water.

Limitations

The shellfish area monitoring network is the only network for monitoring saltwater quality. In general, a shellfish area is classified as permanently closed for shellfish harvesting if fecal coliform density in the water exceeds the salubrity standards mentioned above. Shellfish areas may also be closed for socio-economic reasons, however, or because of other types of contamination or concerns about resource availability. Data on gains or losses of shellfish sites do not therefore necessarily reflect bacterial water quality alone. An area may be closed in the summer, when fecal coliform input is high, and open the rest of the year (conditionally open).

A shellfish area may also be closed because deleterious and toxic substances are present in large enough quantities for consumption of shellfish from the area to be dangerous. In addition, a toxin from the microscopic alga *Alexandrium excavatum* can also contaminate shellfish in some areas. This toxin can cause paralytic poisoning in those who consume the contaminated shellfish. In the St. Lawrence Estuary and Gulf, Softshell clams and Blue mussels are the main carriers of paralytic shellfish poisoning. For a shellfish area to be open, the concentration of this toxin must not exceed 80 µg/100 g of shellfish flesh, and no other shellfish neurotoxins may be present in detectable concentrations.

Problem Or Risk-Prone Sector

Results

Exceedances of fecal coliform standards at sampling stations in 1992 in the four main shellfish harvesting regions are shown in Figure 3.2.6. Of the 2468 sampling stations, 42% are located along the Côte Nord, 33% in Gaspésie, 17% in the Bas-Saint-Laurent and 8% around the Îles de la Madeleine. Results show that water quality around the Îles de la Madeleine and along the Côte-Nord is better than in the Gaspé and in the Bas-Saint-Laurent.

The most contaminated region is the Gaspé, where the spatial median exceeded 14 MPN at 32% of the stations. In addition, at least 10% of the samples exceeded the standard of 43 MPN at 470 of the Gaspésie sampling stations. The Bas-Saint-Laurent comes next, with a spatial median exceeding 14 MPN at 31% of the sampling stations and 204 stations where at least 10% of the samples exceeded the standard of 43 MPN. Along the Côte-Nord, spatial median exceeded 14 MPN at 3% of the stations and at least 10% of the samples

FIGURE 3.2.6
Shellfish harvesting: Exceedances of fecal coliform criteria in various regions of the St. Lawrence (1992)

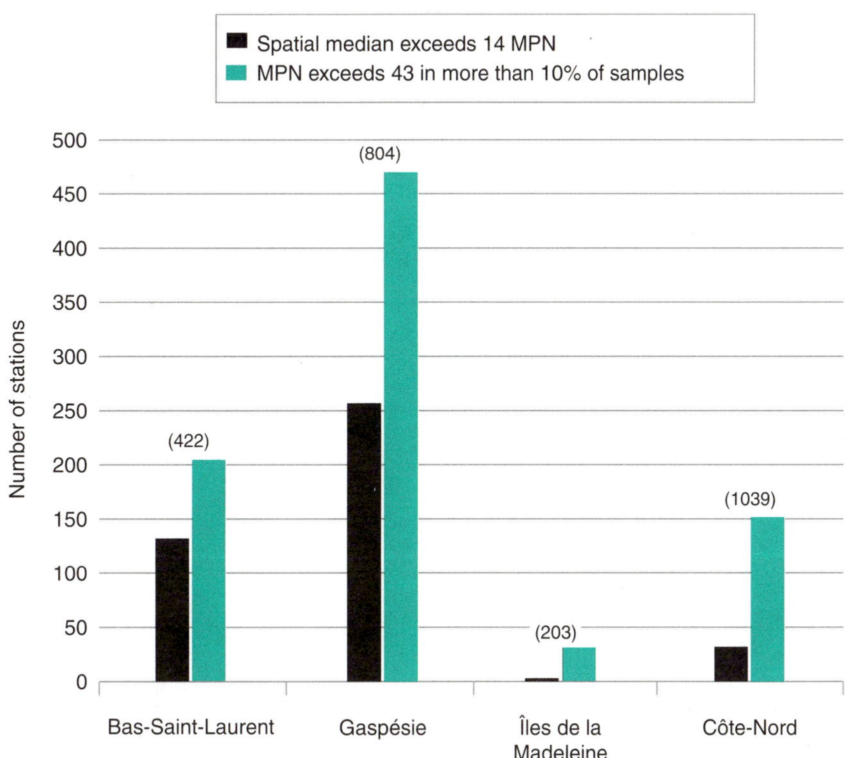

Note: Figures in brackets indicate number of stations sampled in the region.

Source: Based on data from the Environmental Protection Branch, 1992.

exceeded the standard of 43 MPN at 152 stations. The Îles de la Madeleine is the least contaminated region; spatial median exceeded 14 MPN at 2% of the sampling stations and at least 10% of the samples exceeded the standard of 43 MPN at 31 stations.

Contamination indicated by exceedance of fecal coliform standards led to the closing of 85 of the 176 shellfish areas in the four regions. As mentioned (see Limitations above), an area may be closed for reasons of contamination other than bacterial. In 1992, the percentage of closed shellfish areas was highest in the Bas-Saint-Laurent region (26 of 34 shellfish areas, or 76%, were closed). The Gaspé region came next (40 of 58 shellfish areas, or 69%, were closed), followed by the Côte-Nord (14 of 58 shellfish areas, or 24%, were closed) and Îles de la Madeleine (5 of 26 shellfish areas, or 19%, were closed). The closing of these shellfish areas was brought about largely by dense population, extensive farmland in the regions, and riverside homes with outdated sumps and septic tanks.

When bacteriological results exceed standards during a particular sampling campaign at a number of stations, precipitation or an increase in local population (tourists) is assumed. This leads to an increase in contamination and hence closing of the area. This problem arises mainly during snow melt, when heavy rains cause leaching of soil, and during the summer, when tourists invade the banks of the St. Lawrence. This is why many areas are temporarily closed from June to September.

ST. LAWRENCE WATER QUALITY

Main Findings

- Water quality for the uses investigated for this report has improved over the last 15 years because of reduced toxic inputs and an increase in the proportion of riverside populations served by water treatment facilities. Exceedances of quality criteria for the four uses studied (aquatic life, direct human consumption, primary-contact recreation, and shellfish harvesting) were nonetheless found.

Relative Importance

- The water quality of the St. Lawrence is both an influential and an influenced characteristic. It ranks third among the eight most influential characteristics and fourth among the seven most-influenced characteristics.

Recommendations on Monitoring

- To adequately assess fresh water quality in the St. Lawrence, each of its five main water masses must be monitored; that is, the water masses of the Great Lakes, the Ottawa River, the Great Lakes–south shore mix, the Ottawa River–north shore mix and the Quebec City region.
- An index based on the cumulative loss of water uses with five or six classes or categories would help to provide a more complete assessment of water quality in general.

Aquatic life

- The data collection network should be extended downstream of Île d'Orléans so salt water quality for this use can be assessed and major contaminant sources (municipalities, riverside industrial plants and tributaries) can be taken into account.

Direct human consumption

- Existing indicators offer a good assessment of raw drinking water quality. These indicators should be incorporated into a new index based on the cumulative loss of water uses.

Primary-contact recreation

- Water quality data should be available at certain key periods of the year for all stretches of the St. Lawrence, and for the water masses of tributaries where bacterial contamination has been a problem. The indicator used should be considered when developing a new index based on the cumulative loss of water uses.

Shellfish harvesting

- The existing index does not allow clear conclusions to be drawn on salt water quality for different uses. A salt water quality monitoring network should be developed so trends can be tracked using indexes based on bacteriological and physico-chemical factors, such as those developed for fresh water. These indexes should also be incorporated into a new index based on the cumulative loss of water uses.

3.3

TRIBUTARY WATER QUALITY

St. Lawrence Centre, D. Labonté

Background

The interim report of the Comité d'étude sur le fleuve Saint-Laurent (1978) mentions the poor water quality at the mouths of 25 of the 60 tributaries qualitatively assessed at the time. Today, the relative contributions of different sources of toxic inputs to the St. Lawrence can be assessed, and it is estimated that 29% of the toxic input comes from fifty-one tributaries between Cornwall and Quebec City.

These tributaries are in the most industrialized part of the St. Lawrence; that is, the stretch most likely to be affected by toxic substances from diverse sources (industrial plants, agricultural activities and municipalities).

Most of the tributaries studied empty into the Fluvial Section (34) or the Fluvial Estuary (15). Two empty into the Upper Estuary. The Lower Estuary and Gulf were not covered by the study of tributary water quality.

Interpretation

Total annual toxic loads of the following tributaries, in descending order, exceeded 100 000 Chimiotox units in 1991: the Saint-Maurice, Ottawa, Richelieu, Saint-François, Batiscan, Yamaska and Chaudière rivers.

Chimiotox index

Definition

The indicator selected to assess tributary water quality is the Chimiotox index, an index developed to assess reductions in toxic substances (the 124 substances most likely to be found in the St. Lawrence) discharged in effluent from the 50 industrial plants targeted by the St. Lawrence Action Plan. It is thus a theoretical indicator that can be used to track changes in toxic discharges to the St. Lawrence from a specific source.

The Chimiotox index was subsequently used to assess toxic inputs (organic, inorganic and total) from tributaries. Measured at tributary mouths, Chimiotox indexes provide an estimate of tributary contribution to contamination of the St. Lawrence. Among the contaminants measured are five classes of organic compounds (PAHs, PCBs, DDT, BHC and chlordane), five organic substances (atrazine, diazinon, hexachlorobenzene, tetrachlorophenol and pentachlorophenol) and nine metals (cadmium, cobalt, chromium, copper, iron, manganese, nickel, lead and zinc).

Limitations

A Chimiotox index was calculated for only 51 of the 244 tributaries of the St. Lawrence. The index calculated for the 51 tributaries in 1991 was 7.65 million toxic units. The 20 tributaries listed in Figure 3.3.1 account for 97% of this index; that is, 7.43 million units. Nine of these twenty tributaries empty into the main course of the River, nine empty into the Fluvial Estuary and two empty into the Upper Estuary.

In addition, the Chimiotox index can not be used to determine relative contributions of the diverse contaminant sources associated with the tributaries; that is, industrial, urban and agricultural activities.

Mean annual discharges of these tributaries are also indicated, since they are a main factor affecting Chimiotox indexes.

Problem Or Risk-Prone Sectors

Results

Total toxic load of the 20 tributaries listed in Figure 3.3.1 breaks down as follows: toxic organic load (96%) + toxic inorganic load (4%). As mentioned, toxic organic load is the sum of Chimiotox values of five classes of organic compounds plus five organic substances; toxic inorganic load is the sum of the Chimiotox values of nine metals.

Total toxic loads of seven tributaries exceed 100 000 Chimiotox units; they are, in descending order, the Saint-Maurice, Ottawa, Richelieu, Saint-François, Batiscan, Yamaska and Chaudière rivers. The combined Chimiotox index for all these tributaries is estimated at 6.87 million units. These tributaries are, on the whole, those with the highest mean annual discharges.

The highest toxic organic loads (more than 100 000 units) were found in the Saint-Maurice, Ottawa, Richelieu, Saint-François, Chaudière, Batiscan and Yamaska rivers.

Highest toxic inorganic (heavy metals) loads (more than 5000 units) were found at the mouths of the Ottawa, Saint-Maurice, Richelieu, Saint-François, Batiscan, Yamaska and Bécancour rivers.

FIGURE 3.3.1

Total annual toxic loads (Chimiotox indexes) and mean discharges of the 20 largest tributaries of the St. Lawrence (1991)

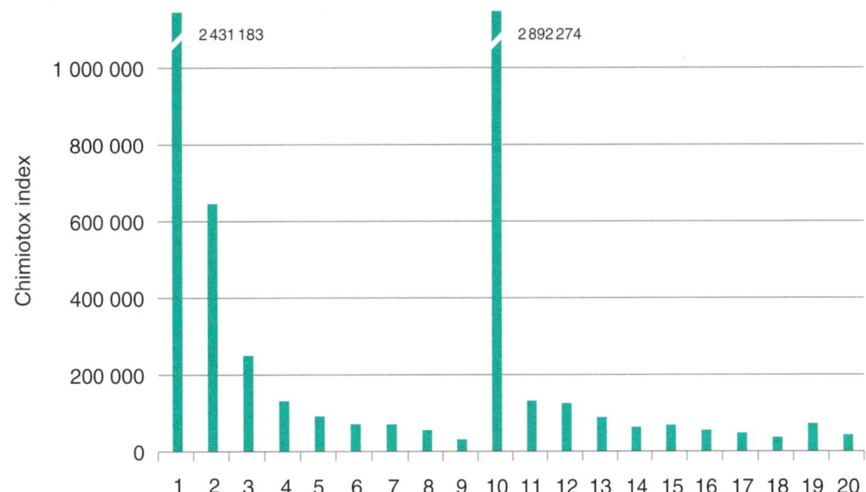

Fluvial Section	Fluvial Estuary	Upper Estuary
1 Ottawa (1152)	10 Saint-Maurice (465)	19 Malbaie (26)
2 Richelieu (208)	11 Batiscan (80)	20 du Loup (10)
3 Saint-François (61)	12 Chaudière (21)	
4 Yamaska (23)	13 Bécancour (22)	
5 Nicolet (10)	14 Jacques-Cartier (38)	
6 du Nord (14)	15 Montmorency (18)	
7 Rouge (<1)	16 Etchemin (6)	
8 L'Assomption (28)	17 Saint-Anne (35)	
9 Saint-Régis (20)	18 Portneuf (13)	

Note:Figures in brackets indicate mean discharges over eight months in m^3/s.

Source: Based on data from Quémerais, 1993.

TRIBUTARY WATER QUALITY

Main Findings

- The waters of the tributaries are loaded with a variety of toxic substances from industrial and agricultural activities and urban wastewater. The contribution of 51 tributaries to the toxic load in the St. Lawrence was measured at their mouths in 1991; 19 contaminants were considered. It is estimated that 29% of toxic inputs to the St. Lawrence come from these tributaries that empty into the St. Lawrence between Cornwall and Quebec City.
- The tributaries with the largest mean annual discharges also contribute the largest total toxic loads (organic and inorganic). The tributaries with loads exceeding 100 000 toxic units include, in descending order, the Ottawa, Saint-Maurice, Richelieu, Saint-François, Batiscan, Yamaska and Chaudière rivers. It is impossible to tell whether toxic loads have decreased or increased since 1977 for lack of adequate data.
- Twenty tributaries (about 40% of the 51 investigated) contribute 97% of the total toxic load discharged to the St. Lawrence by the tributaries. Organics constitute 96% of this toxic load, testifying to the important contribution of agricultural activities to contamination of the St. Lawrence.

Relative Importance

▲ Tributary water quality ranks second in importance among the eight most influential characteristics for the St. Lawrence as a whole.

Recommendation on Monitoring

- Periodic measurement of toxic inputs at the mouths of the tributaries with the highest Chimiotox indexes will increase our knowledge of this characteristic and make it possible to track changes in Chimiotox indexes over time. The study area should be extended to include the Lower Estuary and Gulf.

3.4

BIODIVERSITY

Background

Biodiversity is a complex concept; it draws on the notions of variety and genetic variability in species, ecosystems, processes and functions.

The measurements available have their weaknesses and generally reflect only one of the many aspects of biodiversity, an integrative concept of the quality of living communities in an ecosystem. This assessment addresses two aspects of biodiversity: the precarious situation of certain animal and plant species, and the pressures which the introduction of exotic species exerts on species native to the St. Lawrence.

An information supplement on Black ducks and Mallards in northeastern North America provides additional material on genetic variability.

Biodiversity is also closely linked to the condition of biological resources in terms of abundance and contamination, and to natural environments and protected species. Given current knowledge, and to facilitate discussion of the material, the three characteristics will be dealt with separately.

Interpretation

First, 246 plant species (19% of the vascular plants) and 32 animal species (9% of the animal species) associated with the St. Lawrence are considered Action Plan priority species. The number of these species designated as vulnerable, threatened or endangered has continued to increase, rising from 8 to 20 species between 1984 and 1993. In the same period, the status of two bird species, the Piping plover and Loggerhead shrike, deteriorated from "threatened" to "endangered."

Second, while the full impact of the introduction of exotic species is difficult to quantify, it is a major concern. Exotic species are found throughout the St. Lawrence, depending on their adaptation to fresh or salt water. One example is the Zebra mussel. First sighted in 1989, it has spread throughout almost all the freshwater sections of the St. Lawrence. The impact of changes caused by Zebra mussel colonization on native species is not yet fully known.

Number of animal and plant species at risk

Definition

The St. Lawrence Action Plan has highlighted the precarious situation of certain plant and animal species in the St. Lawrence corridor by designating them priority species. These species show evidence of problems associated with habitat or the state of their populations, among other things. They may also have commercial value. Some of these species have also been assigned a status by the Committee on the Status of Endangered Wildlife in Canada (COSEWIC). The three main statuses assigned are "vulnerable," "threatened" and "endangered."

The fact that both animal and plant species have been designated as priority species indicates that it will require considerable effort to maintain the diversity and integrity of plant and animal communities in the St. Lawrence ecosystem.

Limitations

The length of the list reflects advances in knowledge and, to a certain extent, the degree of interest in priority species. Their identification involves learning more about the species, their populations and their habitats to gain a better understanding of the risks associated with decreased biological diversity in the St. Lawrence.

Problem Or Risk-Prone Sectors

Results

Nineteen percent of the vascular plants associated with the St. Lawrence had priority status under the St. Lawrence Action Plan in 1993. These 246 species were made up of 9 trees, 14 shrubs and 223 herbaceous plants (see Figure 3.4.1). Thirty-two or 9% of wildlife species associated with the River had priority status under the Action Plan: 11 fish, 2 amphibians, 5 reptiles, 11 birds and 3 marine mammals. Per wildlife group, this breaks down as, fish (6%), amphibians (13%), reptiles (36%), birds (10%) and marine mammals (15%).

These priority species are found mainly in the Fluvial Section, especially at Lake Saint-Pierre, and in the Fluvial Estuary. Eleven fish species are found in these sectors, along with two species of amphibian, five reptiles and eleven birds. Lake Saint-Louis alone accounts for more than 40 priority vascular plant species. All three marine mammals at risk are associated with the Upper Estuary, Lower Estuary and Gulf.

FIGURE 3.4.1

Number of priority animal and plant species under the St. Lawrence Action Plan and associated with the St. Lawrence River (1993)

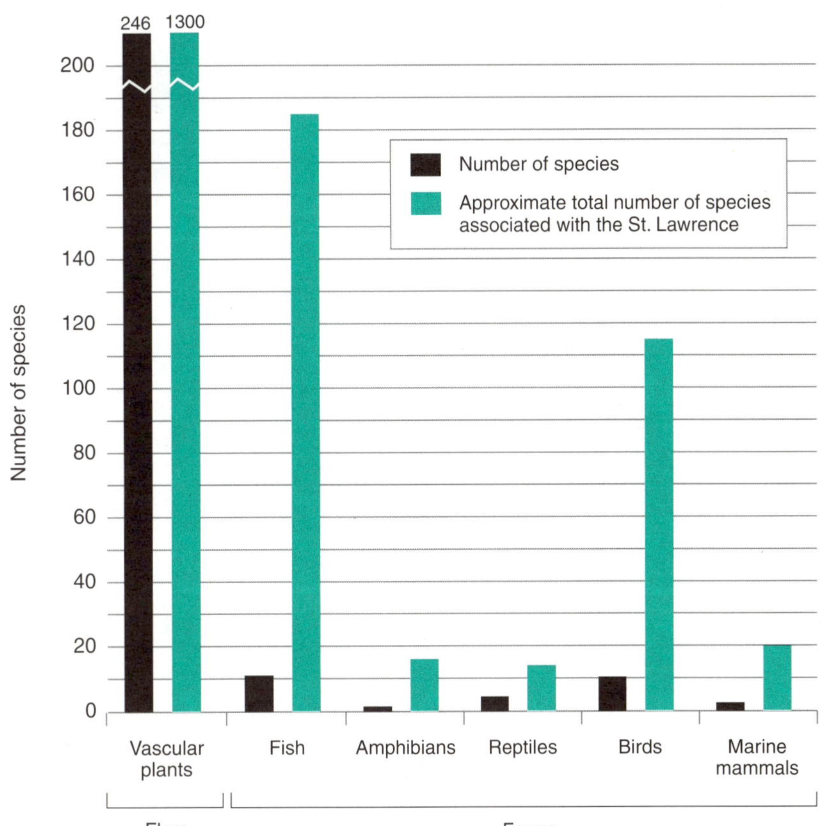

Sources: Based on data from Lavoie, 1992; Groupe de travail sur les espèces de faune et de flore prioritaires du couloir du Saint-Laurent, 1993; MENVIQ, 1992a; Ducharme et al., 1992; Ghanimé et al., 1990; Gratton and Dubreuil, 1990; Kingsley, 1995; Hammill, 1995.

In 1993, there were 20 species with both official COSEWIC-designated status and priority status under the St. Lawrence Action Plan (see Figure 3.4.2). These included nine vascular plants, two fish, one reptile, six birds and two marine mammals. Three species of birds have had COSEWIC status since 1978: the Peregrine falcon, Piping plover and Loggerhead shrike. Of the 20 designated priority species, nine were classified as vulnerable, seven as threatened and four as endangered.

Between 1984 and 1993, the number of species assigned a status by COSEWIC rose from 8 to 20; it was 13 in 1987. During this period, the number of vulnerable species rose from four to nine, the number of threatened species from two to five and the number of endangered species from two to four. The status of the Piping plover was revised in 1985 and that of the Loggerhead shrike in 1991. Previously considered threatened, both species are now on the endangered list.

FIGURE 3.4.2

Changes in COSEWIC-designated status for priority species under the St. Lawrence Action Plan (1984-1993)

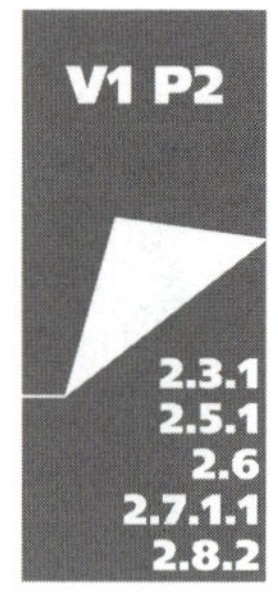

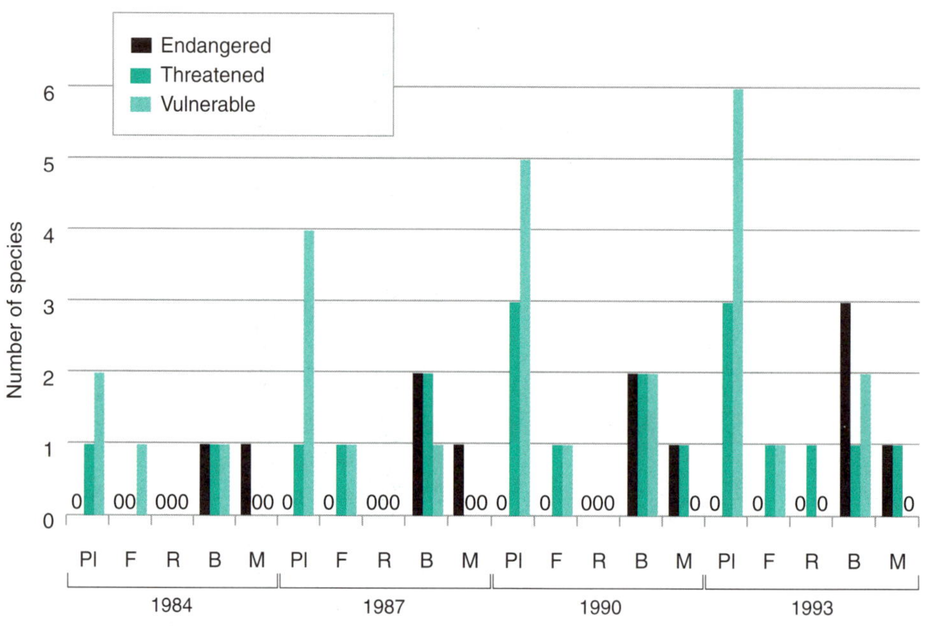

Note: Pl stands for vascular plants, F for fish, R for reptiles, B for birds, M for marine mammals.

Source: Based on data from COSEWIC, 1993.

Density of Zebra mussels on navigation buoys

Definition

Exotic species – whether introduced deliberately or accidentally – can encroach upon native species, disturbing food webs and competitive relationships that have shaped community structures over many years. The degree of introduction (up to invasion) of an exotic species into the fluvial ecosystem can be assessed by measuring changes in the species' density in the environment.

Data from 1990 to 1992 on annual colonization by Zebra mussels in one part of the St. Lawrence are shown in Figure 3.4.3. The Zebra mussel was introduced into the St. Lawrence through ship ballast water. Mussel density per square metre was measured on navigation buoys in the River corridor, from Lake Saint-François to Sault-au-Cochon. This measurement is an accurate representation of annual colonization by mussels because the buoys are cleaned each winter.

Limitations

The indicator used reflects annual juvenile recruitment. It can not be used to assess established communities.

Problem Or Risk–Prone Sectors

Results

From 1990 to 1991, annual colonization increased significantly in all parts of the River corridor, except in Lake Saint-François, where it dropped sharply. From 1991 to 1992, annual colonization decreased in most sectors, except in Lake Saint-François and the Beauharnois canal, where year-to-year differences in density were not significant.

The Zebra mussel was sighted in 1989 and has spread throughout the freshwater section of the St. Lawrence.

FIGURE 3.4.3

Density of Zebra mussels attached to navigation buoys, from Lake Saint-François to Sault-au-Cochon (1990-1992)

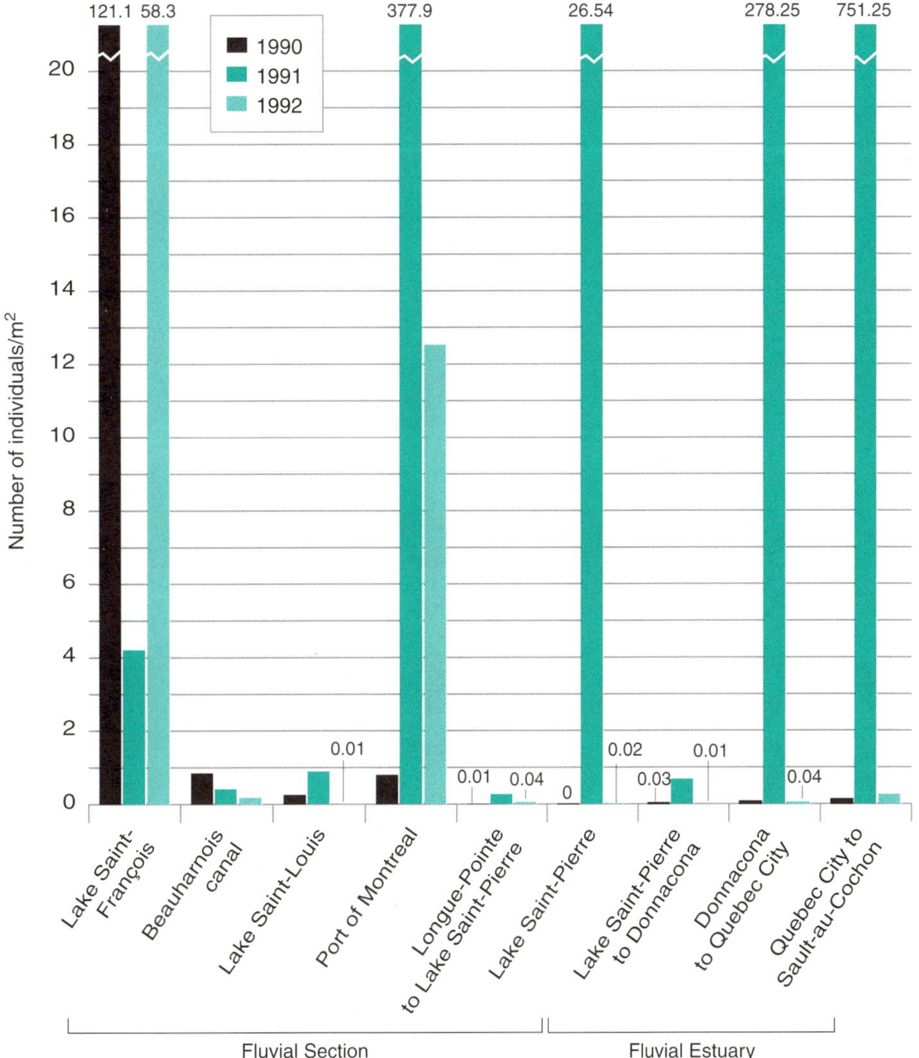

Source: Based on data from Lapierre and Fontaine, 1994.

Information Supplement

BLACK DUCKS AND MALLARDS IN NORTHEASTERN NORTH AMERICA

Winter inventories of waterfowl conducted in the United States since 1955 show a marked decline in the number of Black ducks observed in the Atlantic flyway, which passes over the St. Lawrence River. Between 1955 and 1988, numbers dropped by 44%, from an average of 398 117 in the 1955-1960 period to 222 375 in the 1981-1988 period.

There may be several explanations for the decline of the Black duck in relation to the Mallard. The Black duck has lost many habitats to urban development, while the Mallard seems to adapt more easily to habitats altered by human intervention. Introduction of the Mallard into northeastern North America, where the Black duck dominated, has led to interbreeding between the two species, resulting in fertile offspring that have the dominant genetic characteristics of the Mallard.

Has recent colonization of northeastern North America by Mallards affected present Black duck populations? Before 1900, Mallards were seldom sighted in northeastern North America. In the 1920s, the Mallard started to become more and more common in the Northeast. The expansion of its range into this region between 1900 and 1950 was probably a result of human modification of the environment, which created habitats suitable for Mallards.

The Black duck and Mallard are genetically similar; the Black duck is probably a direct descendent of the Mallard. The Mallard population in the Northeast was probably segregated from western populations by repeated glaciation during the Pleistocene period. Through natural selection, this population gradually evolved into the species now called the Black duck. The duration of the separation was not great enough, however, for the Black duck to evolve into a completely different species from the Mallard. The two species share similarities in feeding mechanisms, courtship display patterns and size and structure; only the habitat occupied differs.

Will the Black duck become extinct as a species distinct from the Mallard? In areas where the two species are in contact, there seems to be a strong correlation between declining numbers of Black ducks, increasing hybridization between the two species and the increase in the number of Mallards in northeastern North America since the early 20th century. Other factors, including loss of habitat and overhunting, may also have played a role in the decline of the Black duck in northeastern North America. If the Black duck's habitat is altered or disappears due to human modifications of the environment, this species may disappear with it.

Source: Based on data from Heusmann, 1974; Rusch et al., 1989.

BIODIVERSITY

Main Findings

- There are about 1300 vascular plant species, 185 fish species, 115 bird species, 16 amphibian species, 14 reptile species, and 20 mammal species associated with the St. Lawrence.
- The number of animal and plant species at risk has increased steadily since 1983. Between 1984 and 1993, the number of animal species designated vulnerable, threatened or endangered rose from eight to twenty. Within the same period, the number of endangered species increased from two to four. The St. Lawrence beluga is still designated an endangered species.
- In 1993, 32 animal species and 246 vascular plant species, or about 9% of the animal species and 19% of the vascular plant species associated with the St. Lawrence, were assigned priority status under the St. Lawrence Action Plan. Lake Saint-Louis alone is home to more than 40 of the 246 vascular plant species with priority status.
- The Zebra mussel is widely distributed throughout the freshwater section of the St. Lawrence, yet the impact of this colonization is not fully known.

Relative Importance

- Due to the lack of knowledge about biodiversity, this characteristic ranked last of the seven most-influenced characteristics for the River as a whole.

Recommendations on Monitoring

- Monitoring biodiversity involves many unknowns, especially where plant and animal communities, and their place and role within the ecosystem, are concerned. To increase scientific knowledge, biodiversity should be measured in several different ways and complex indexes of biological integrity should be developed. It will be possible to use this characteristic to better assess the state of the St. Lawrence by combining information on the different aspects of biodiversity. An effort should also be made to study the river ecosystem as a whole, not just on a species-by-species basis. Among other things, a better understanding should be developed of the extent to which biodiversity is influenced by the condition of biological resources and shoreline modifications.

- Changes in the number of priority species enables only one aspect of biodiversity to be assessed. New indicators that take other aspects into account should be developed. Changes in the status of specific species indicate changes in their environments.
- The monitoring of introduced plant and animal species provides basic information only; additional information is required for a better understanding of their impact on the environment.

3.5

NATURAL ENVIRONMENTS AND PROTECTED SPECIES

MLCP, M. Beaudoin

Background

Setting aside protected natural environments may be the only way to ensure the survival of several resources that are typical of the St. Lawrence. Most of the time, these natural environments are shoreline wetlands. But these habitats also include marine areas, cliffs, islands, shallows, tidal flats and reefs, all essential to the harmonious functioning of the fluvial ecosystem. The protection of natural environments should be focused on the richest and most complex communities that contain the rarest species, without neglecting environments characterized by extreme ecological conditions.

One of the best ways to conserve natural environments is to give them legislative protection. Today, some 290 200 hectares (ha) adjoining all four parts of the St. Lawrence have received legal protection from federal, provincial or municipal governments. The protection afforded varies widely, depending on the legal status assigned; such protection is an important factor in ensuring long-term species survival. The purchase of natural environments (without legal protected status) by governments or non-governmental organizations is another way in which these natural environments can be conserved. Along the St. Lawrence, 8748 ha have no legal status but are protected in this manner.

Interpretation

Analysis of the data shows an inequity in the location of protected areas and the distribution of animal and plant species at risk. The most-threatened species and wildlife communities associated with the St. Lawrence or its wetlands are concentrated in the two upriver sections, the main course of the River and the Fluvial Estuary, particularly in Lake Saint-Pierre. Protected natural environments account for only relatively limited surface areas in these two sectors. When one considers the distribution of rare and threatened plant species and communities along the St. Lawrence, it becomes apparent that the Fluvial Section and the Fluvial Estuary are in greatest need of protected habitats.

Surface area of protected natural environments, by category

Definition

Hectares of protected natural environments are broken down into the five following categories:

Category I – National parks, provincial conservation and recreation parks, and ecological reserves under provincial jurisdiction covered by specific legislation, with allocation of material resources to increase knowledge of the environment and enforce regulations.

Category II – National wildlife areas and migratory bird sanctuaries (under federal jurisdiction), the wildlife preserve (provincial jurisdiction) and certain private sites providing excellent assurance of protection of biological resources.

Category III – Staging areas for migratory birds, municipal regional parks, provincial wildlife sanctuaries and some nature centres in natural environments.

In practice, few resources are available for enhancement (species inventories have not been compiled, while public access is variable and occasionally conflicts with conservation goals).

Category IV – Islands that belong to private conservation groups, and protected landscape areas.

Compared to sites in the first three categories, these sites are less strictly protected; resources for enhancement and protection vary, but are generally few.

Category V – Certain types of wildlife habitats designated and mapped by the Ministère de l'Environnement et de la Faune because of their essential role in the survival of certain wildlife species and to provide special protection under the *Act respecting the Conservation and Enhancement of Wildlife*.

These include heronries, bird colonies, and areas where aquatic birds congregate.

Data on surface areas (ha) of protected habitats by category and hydrographic region in 1992 are shown in Figure 3.5.1. The number of hectares by hydrographic region can not be determined at this time for Category V. Figures for Category V are based on the whole of Quebec, for the three types of wildlife habitats most likely to be found along the St. Lawrence.

Limitations

Using the number of hectares of protected natural environments as an indicator of the state of this characteristic gives only a partial picture of spaces and excludes species. In addition, it is a single measurement (the only one available at present) and does not provide enough data to draw clear conclusions about natural environments: it is not weighted for the area's status, type, diversity, or the quality and abundance of protected biological resources. It does, however, allow the distribution of these environments and the distribution of species and communities at risk to be compared.

Results

Problem Or Risk-Prone Sectors

FLUVIAL SECTION | FLUVIAL ESTUARY

The distribution of protected areas varies greatly from one section of the River to another – differences not made any clearer by unequal shoreline development in the four sections. Looking at categories I through V, we see that the Fluvial Estuary has almost no protected areas (886 ha) and the Fluvial Section has very few (20 653 ha). In the Upper Estuary, the surface area increases to 35 465 ha (excluding the purely marine component of the Saguenay Marine Park). If the Saguenay Marine Park is included, this figure rises to 149 265 ha. The terrestrial component of the Saguenay park (28 360 ha) adds considerably to the relative weight of this section. In the Lower Estuary and Gulf, the surface area is 138 255 ha of shoreline. The Anticosti Island Wildlife Preserve (53 900 ha), Forillon National Park (24 040 ha) and Mingan Archipelago National Park Reserve (15 000 ha) are the largest protected spaces in this river section.

FIGURE 3.5.1

Surface area of protected natural environments, by category and river section (1992)

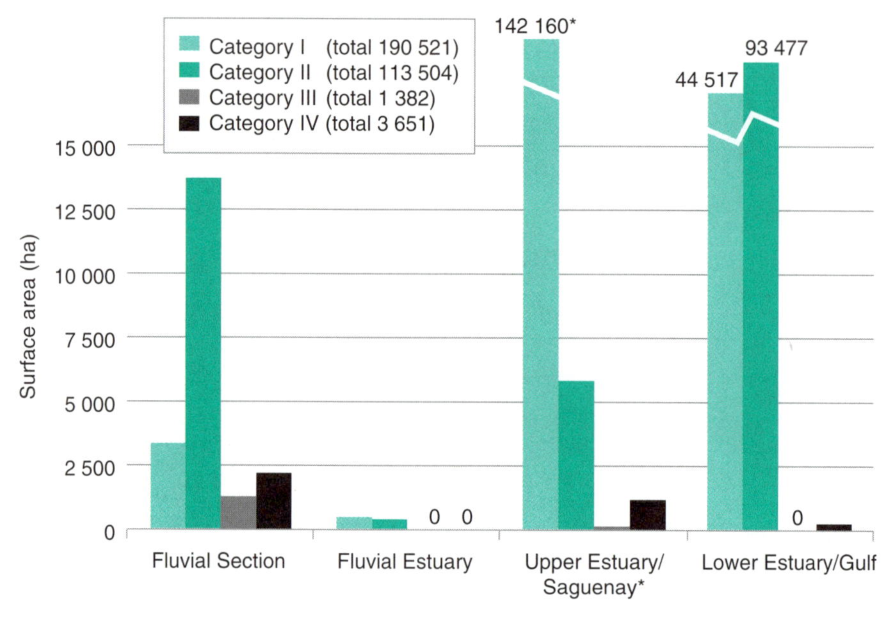

*Includes the Saguenay Marine Park.

Note: Category V total is 35 500 ha for all of Quebec.
Total for all categories is 344 558 ha.

Several protected natural environments have a dual status; the number of hectares may therefore have been overestimated.

Source: Based on data from Boucher, 1992.

NATURAL ENVIRONMENTS AND PROTECTED SPECIES

Main Findings

- The survival of a number of resources that are typical of the St. Lawrence may depend on the protection of natural environments. Analysis of 1992 data shows an inconsistency between the location of protected areas and the distribution of animal and plant species at risk. The most threatened animal species and communities are in the two upriver sections of the St. Lawrence. The largest concentration of fish spawning grounds are in Lake Saint-Pierre and the Sorel islands. Looking at the distribution of rare and threatened plant species and communities along the St. Lawrence, it is clear that the main course of the River and Fluvial Estuary show the greatest need for protected areas. Yet there are relatively few protected natural environments in the former (20 653 ha). In the Fluvial Estuary, such areas are almost nonexistent (886 ha). There are still unprotected natural environments with good potential in the littoral fringe of this sector (for example, from Saint-Augustin-de-Desmaures to Sainte-Anne-de-la-Pérade). Surface area may vary depending on the nature of the space protected and its role in the life cycles of different species.

Relative Importance

▲ This characteristic ranks fourth of the eight most influential characteristics for the St. Lawrence as a whole.

Recommendations on Monitoring

- A measurement must be developed that weights surface areas by the status assigned to an area, based on the type, diversity, quality and abundance of biological resources protected.
- A method for measuring the productivity of protected natural environments must also be developed to improve monitoring of this characteristic.

3.6

CONDITION OF BIOLOGICAL RESOURCES

Canadian Wildlife Service

Background

Two aspects were considered in assessing the condition of biological resources: abundance and contamination. Species abundance, whether positive or negative, was selected because of its close correlation with organism condition. Abundance is the number of individuals caught or inventoried, or an estimate of total biomass. Contamination is determined by finding the concentration of organic and inorganic contaminants in the flesh, liver or eggs of certain species. When measured simultaneously in several species from different taxons, these two indicators give some indication of the condition of living organisms and their habitat.

For this report, the species used to illustrate the condition of biological resources were selected on the basis of data availability, among other factors. For some species, these data cover a period of more than 20 years. The species chosen are from three taxons (fish, birds and mammals) and live in different habitats (fresh water and salt water). Abundance data are presented for 11 species, and contamination data for 9 species. Data on abundance and contamination of the Northern gannet on Bonaventure Island in the Gaspé cover approximately the same 20-year period.

Interpretation

Abundance varied with species. Several of the 11 species studied had very low abundance. Rainbow smelt, American eel and Atlantic tomcod have continued to decline for a number of years; catches of Rainbow smelt and American eel have dropped off significantly since 1986. Cod stocks have declined steadily since 1985. From 1982 to 1992, the population of Beluga whale, an endangered species, remained fairly steady at 500 members. Abundance of Lake whitefish is still low, but has gradually increased since 1985. Walleye is the most abundant of the fish species selected (the catch increased gradually after 1987 and levels have remained steady since 1989). The Northern gannet population increased gradually from 1969 to 1989 (by 17%, despite a marked decrease in 1976). The population of Greater snow geese is very high and has registered the largest increase (numbers rose fivefold between 1969 and 1992).

In general, it has been shown that toxic contamination of organisms has decreased. This is so in the case of DDE (a derivative of DDT) and polychlorinated biphenyls (PCBs) in Northern gannet eggs from Bonaventure Island, as well as for mercury and cadmium in beluga livers.

The steady decrease in mercury levels in the flesh of Saguenay shrimp is another illustration of the gradual reduction in contamination after certain sources of pollutants were reduced or eliminated. Data from 1972-1992 on PCB levels in eggs of the Double-crested cormorant, a fish-eating bird in the Great Lakes and Lower Estuary, are also examples of this decreasing trend (see information supplement entitled, "PCB levels in the eggs of the Double-breasted cormorant on Île aux Pommes").

Abundance of certain species: Number of individuals caught or inventoried and estimated total biomass

Definition

The number of individuals caught or inventoried or the estimated biomass (for various periods), were established for the following: eight fish species – Rainbow smelt (1963-1992), Walleye (1971-1992), Atlantic tomcod (1971-1992), Lake whitefish (1971-1992), American eel (1975-1992), Atlantic cod (1978-1992), redfish (1978-1992), and Greenland halibut (1984-1992); two bird species – Greater snow goose (1969-1992), Northern gannet (six years between 1969 and 1989); and one marine mammal, the St. Lawrence beluga whale (for six years between 1973 and 1990).

Limitations

This indicator only partially reflects the abundance and quality of biological resources since it does not consider aspects such as reference species and physiological condition.

Abundance data for fish other than cod, redfish and Greenland halibut were only collected at two sources: the Quebec Aquarium and the eel ladder at the Moses Saunders dam in Cornwall.

Problem Or Risk-Prone Sectors

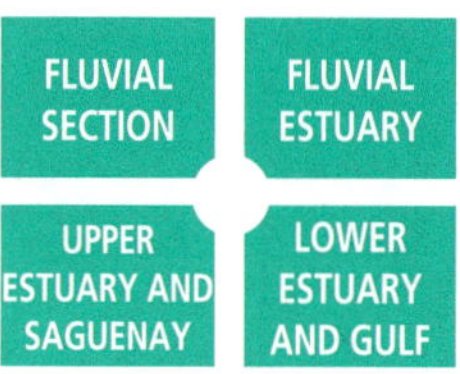

Results

Rainbow smelt are found in several distinct populations: Chaleur Bay, the Saguenay River, on the north shore in the Charlevoix region, and on the south shore of the estuary. In the early 1960s, smelt were abundant and an important resource for the Quebec economy. Since 1963, when 1800 were caught at the Quebec Aquarium, numbers have declined considerably; only 17 were caught in 1992. Yet smelt numbers show considerable year-to-year variation. In the past three decades, its numbers were highest in 1968 (1900 individuals) and lowest in 1991 (7 individuals) (see Figure 3.6.1).

Catches of Walleye increased by 370% between 1971 and 1992, rising from 280 in 1971 to 1315 in 1992. This period was characterized by an alternating two- to four-year period of a drop in stocks, immediately followed by an increase. In the 1988 to 1989 period, the number of Walleye caught increased by 83% and has remained above 1000 since 1989. In the past twenty years, the largest catch was 1315 in 1992, and the lowest was 280 in 1971.

The Atlantic tomcod population in the Upper Estuary, from the eastern tip of Île d'Orléans to Cacouna, has suffered a major decline in the past 20 years. It was still very low in 1992. Between 1971 and 1992, the number of individuals captured dropped by 99%, from 2761 in 1971 to 16 in 1992. In the 20-year period, the smallest catches were four fish in 1987 and five fish in 1991; the largest catch was 7156 fish in 1972. The catch increased considerably from 1977 to 1982, but since 1982 has decreased almost continuously.

The Lake whitefish catch has increased since 1971, from 5 in 1971 to 400 in 1992. The best year was 1990, when 575 fish were caught; the worst years were 1971, with five caught, and 1973 and 1974, with seven.

Counts of young American eels at the Moses Saunders dam in Cornwall since 1975 show that a major decline started in 1986 (see Figure 3.6.2). Between 1975 and 1992, numbers dropped by 99%, from 936 000 in 1975 to 11 533 in 1992. Between 1975 and 1985, the average eel count per year was 888 000; from 1986 to 1992, it fell to 192 000. Since 1975, highest levels counted were in 1983, with 1 293 570, and lowest levels in 1992, with 11 533.

Total biomass of cod aged three years or more reached a peak of 771 000 tonnes in 1983 and has decreased steadily since then. The lowest cod stocks were recorded in 1992 (291 000 t). The sharpest drop occurred between 1986 and 1988, when estimated stocks declined by 22% each year (see Figure 3.6.3).

Redfish biomass fluctuates considerably from year to year. Highest biomass was 211 000 tonnes, in 1979; lowest biomass was 27 000 t, in 1987. Between 1983 and 1986, the average value was 134 000; between 1987 and 1992, it was 49 000. The sharpest drop occurred between 1986 and 1987, when redfish stocks declined 81% in a single year.

FIGURE 3.6.1

Annual catch of four fish species using fixed nets at the Quebec Aquarium (1963-1992)

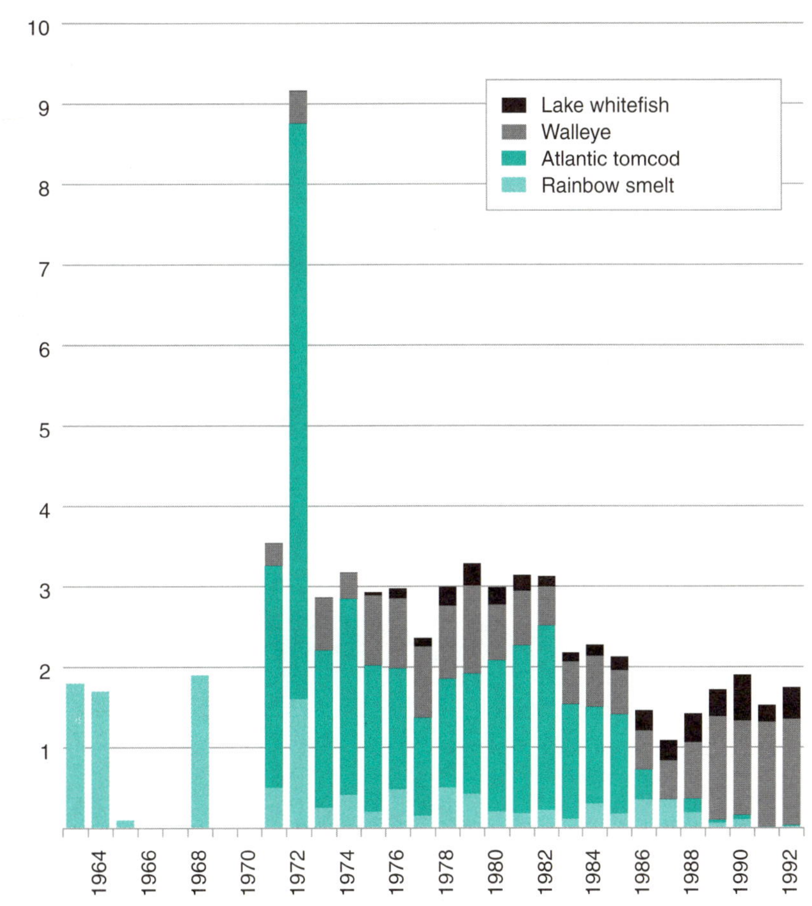

Note: Where no bar appears, the data were unavailable.

Source: Based on data from Pilote, 1993.

Biomass data on Greenland halibut has been available since 1984. Between 1984 and 1992, the highest biomass occurred in 1986 (45 000 t); the lowest occurred in 1990 (12 000 t). Between 1984 and 1992, the average value was 25 000 tonnes.

Data on Greater snow geese show that the spring population increased significantly from 1969 to 1992. Numbers have increased fivefold over more than 20 years (see Figure 3.6.4). As for the Northern gannet, the number of pairs rose from 20 500 in 1969 to 24 000 in 1989. A sharp decrease was recorded in 1976, when only 16 500 pairs were counted. Historical data indicate that the number of nesting pairs declined by almost 25% from 1966 to 1976, due to DDT contamination in the 1970s.

Historical data show the beluga population was 5000-strong in 1885. Since the early 1970s, the population has remained steady at about 500 individuals (see Figure 3.6.5).

FIGURE 3.6.2

Number of young American eels counted at the eel ladder at the Moses Saunders dam in Cornwall (1975-1992)

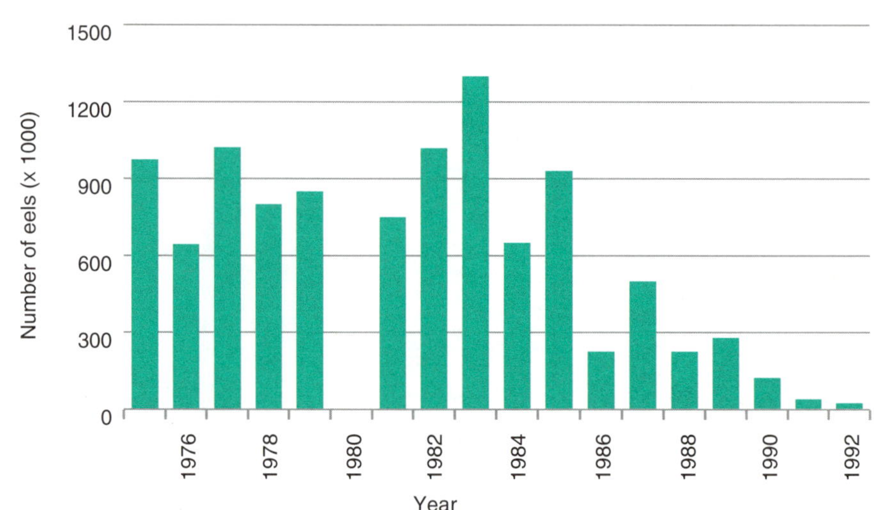

Note: Where no bar appears, the data were unavailable.

Source: Based on data from Eckersley, 1982; Hendrick, 1991; Castonguay et al., 1994a.

FIGURE 3.6.3

Total estimated biomass for cod, redfish and Greenland halibut in the Gulf of St. Lawrence (1978-1992)

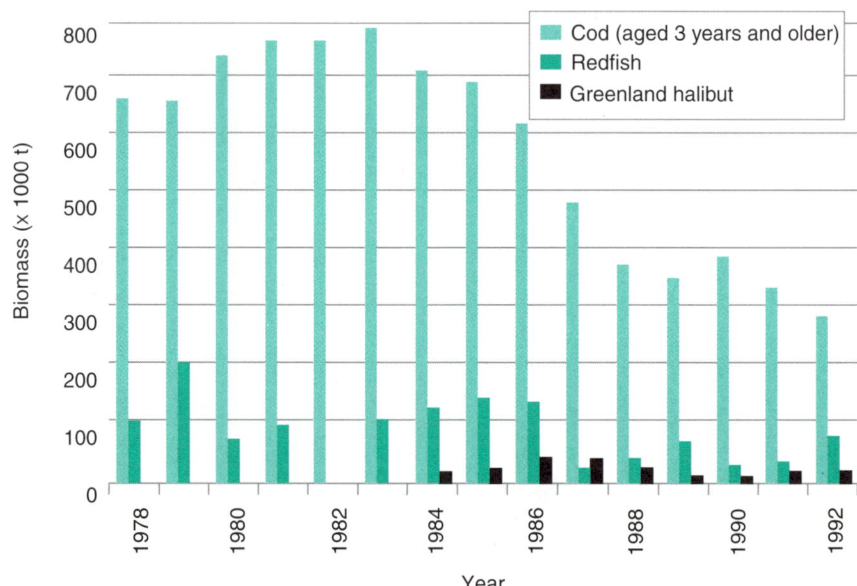

Note: Areas counted – cod (3Pn – 4RS, 4T – 4Vn); redfish (3Pn – 4RS, 4T – 4Vn); Greenland halibut (4R, 4S, 4T).

Where no bar appears, the data were unavailable.

Source: Based on data from Fréchet, 1994; Chouinard, 1994.

FIGURE 3.6.4

Spring population of Greater snow geese in the St. Lawrence Valley and the number of Northern gannet pairs on Bonaventure Island (1969-1992)

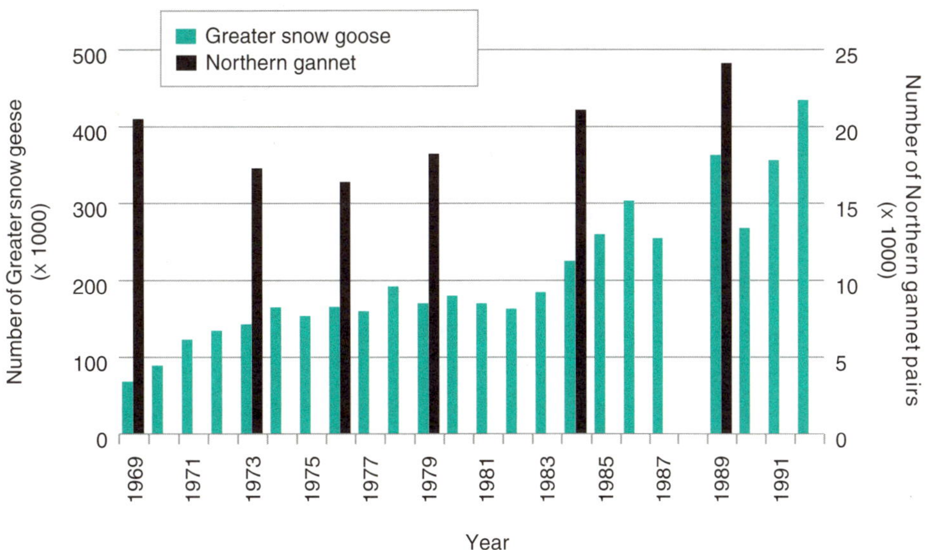

Note: Where no bar appears, the data were unavailable.

Source: Based on data from Chapdelaine et al., 1987; Chapdelaine, 1993 (Northern gannet); Reed, 1993 (Greater snow goose).

FIGURE 3.6.5

Changes over time in the St. Lawrence beluga (*Delphinapterus leucas*) population (1973-1992)

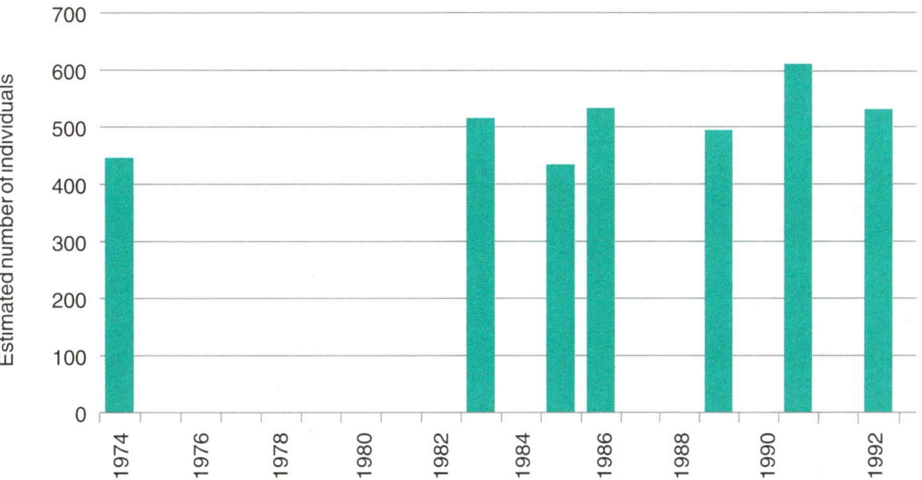

Note: Where no bar appears, the data were unavailable.

Source: Based on data from Kingsley and Hammill, 1991; Sergeant and Hoek, 1988; Kingsley, 1994.

Contamination of certain species: Concentrations (in mg/kg) of various contaminants in flesh, liver or eggs

Definition

Concentrations (in mg/kg) of organic and inorganic contaminants in the flesh, liver or eggs of certain species were measured. The figures show, for the various regions of the St. Lawrence, spatial and/or temporal variations in the levels of several contaminants in the following species: mercury (Hg) and polychlorinated biphenyls (PCBs) – Brown bullhead, Yellow perch, Northern pike and Walleye (1984-1987); mercury (Hg) – cod (1987-1989), shrimp (1970-1989); mirex – American eel (1982 and 1990); hexachlorobenzene (HCB), DDE, polychlorinated biphenyls (PCBs) and dieldrin – Northern gannet (1969-1984); mercury (Hg) and cadmium (Cd) – Beluga whale (1988-1990).

Limitations

The species selected, their place in the ecosystem and the number of contaminants measured have a direct influence on the overall assessment of contaminant levels in the bioresources of the St. Lawrence.

Problem Or Risk-Prone Sectors

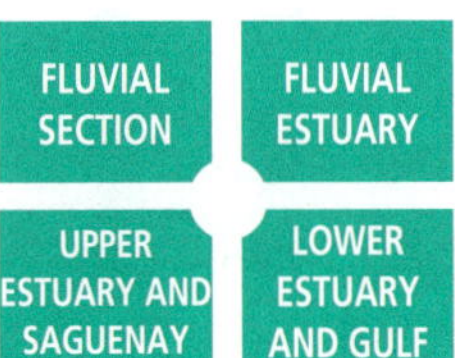

Results

Mercury levels measured in the flesh of adult fish of four species in the fluvial lakes from 1984 to 1987 (see Figure 3.6.6) showed mercury biomagnification. Predatory or piscivorous fish higher up the food chain, such as pike and Walleye, had higher levels of mercury than bottom-dwelling fish such as Yellow perch and Brown bullhead. Marketing guideline levels (0.5 mg/kg) for pike and Walleye were exceeded between 1984 and 1987. The highest mercury levels in Walleye were found in the Ottawa River; mercury levels in Northern pike, Yellow perch and Brown bullhead were highest in Lake Saint-Louis.

During the same period, analysis of the flesh of adult fish showed that PCB levels fluctuated by species and sector examined (see Figure 3.6.7). At no time did levels exceed marketing guidelines. The highest PCB levels were found in Lake Saint-Pierre for Walleye and Brown bullhead, and lakes Saint-Louis and Saint-François for Northern pike and Yellow perch.

Between 1987 and 1989, mercury levels in cod were significantly higher in the flesh of individuals captured in the Saguenay River (sometimes exceeding marketing guidelines), compared to the very low levels in cod caught off the Côte-Nord or in the Gulf (see Figure 3.6.8).

FIGURE 3.6.6

Spatial variations in mercury levels in the flesh of some adult fish in the St. Lawrence (1984-1987)

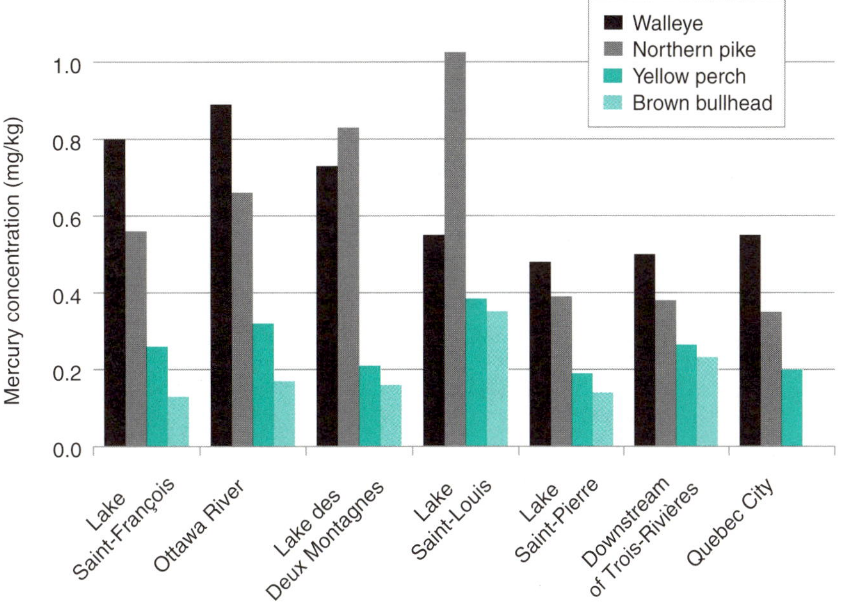

Note: Combined results for each sector, average values in flesh for the 1984-1987 period.

Where no bar appears, the data were unavailable.

Source: Based on data from Legendre and Sloterdijk, 1988.

FIGURE 3.6.7

Spatial variations in PCB levels in the flesh of some adult fish in the St. Lawrence (1983-1986)

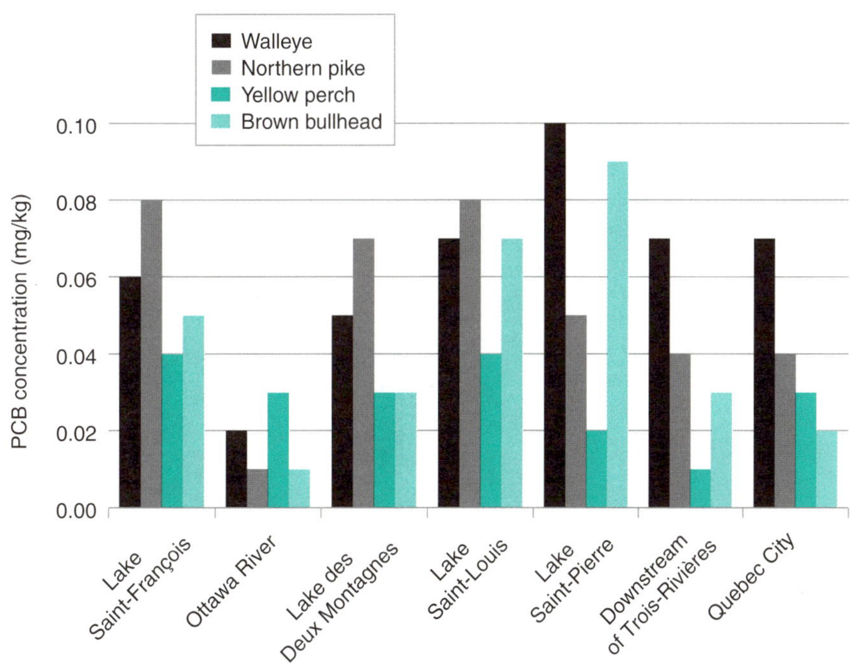

Note: Combined results for each sector, average values in flesh for the 1983-1986 period.

Source: Based on data from Legendre and Sloterdijk, 1988.

FIGURE 3.6.8
Spatial variations in mercury levels in the flesh of St. Lawrence cod (1987-1989)

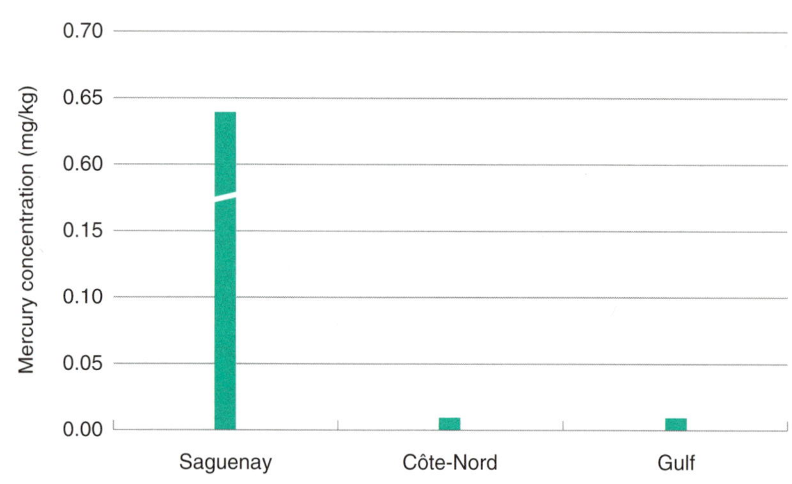

Source: Based on data from De Ladurantaye et al., 1990.

Eels caught near Kamouraska (see Figure 3.6.9) were contaminated by mirex, though levels of mirex have declined considerably since the early 1980s and were much lower in 1990 (0.025 mg/kg) than in 1982 (0.1 mg/kg). Fifty-two percent of eels had mirex levels exceeding the marketing guideline of 0.1 mg/kg in 1982; it was 29% in 1990. Chemical contamination (including mirex) of eels correlates with sources of pollution in the regions where eels grow to maturity. At Kamouraska, it was observed that eels nearing the end of their migration period tended to exhibit higher levels of contamination. Since eels are migratory, knowledge of the migratory patterns of eels coming from different places is very important to data analysis.

FIGURE 3.6.9
Temporal variations in mirex levels in the flesh of American eels at Kamouraska (1982 and 1990)

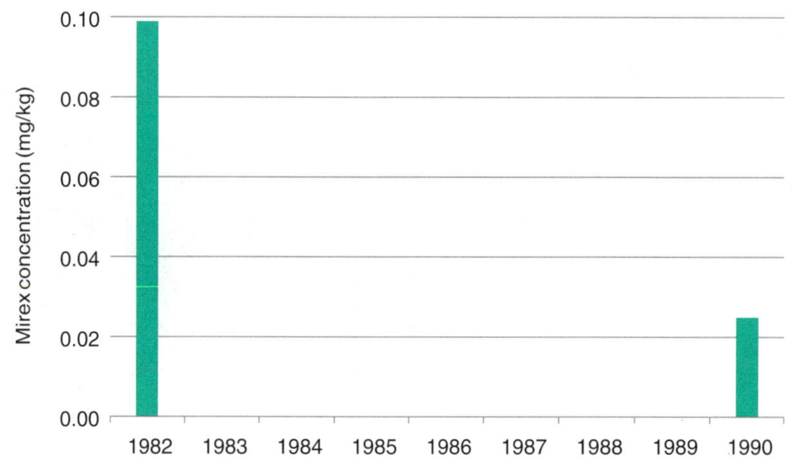

Note: Where no bar appears, the data were unavailable.

Source: Based on data from Hodson et al., 1992.

In 1970, mercury levels in the flesh of Saguenay shrimp were higher than 9 mg/kg (see Figure 3.6.10). Almost 20 years later, levels were in the area of 0.4 mg/kg, or slightly above marketing guidelines.

FIGURE 3.6.10

Changes over time in mercury levels in the flesh of Saguenay shrimp (1970-1989)

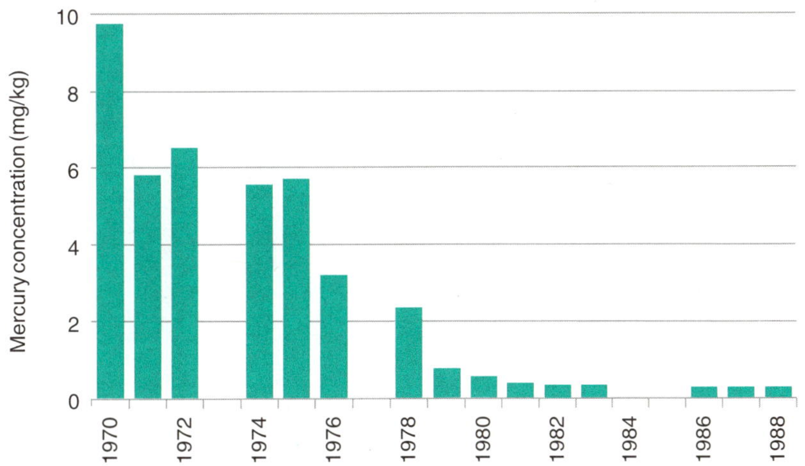

Note: Where no bar appears, the data were unavailable.

Source: Based on data from De Ladurantaye et al., 1990.

As shown in Figure 3.6.11, analysis of fresh eggs gathered between 1969 and 1984 from the Northern gannet colony on Bonaventure Island showed that levels of various organic contaminants gradually decreased. Contamination by DDE (a DDT derivative) dropped after DDT was banned in the 1970s.

FIGURE 3.6.11

Changes over time in concentrations of DDE and PCBs in the eggs of Northern gannets on Bonaventure Island (1969-1989)

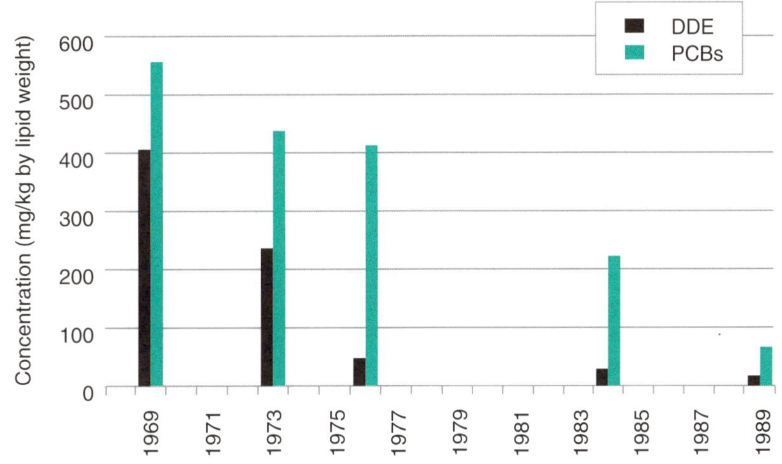

Source: Based on data from Chapdelaine et al, 1987; Chapdelaine, 1993.

Levels of mercury and cadmium in the livers of St. Lawrence belugas from 1988 to 1990 indicate that mercury and cadmium tend to accumulate with age (see Figure 3.6.12); this is primarily significant for mercury. Mercury levels in the livers of St. Lawrence belugas are higher than in most cetaceans in the Northern Hemisphere. On the other hand, the tissues of St. Lawrence belugas contain less cadmium than those of Arctic belugas and most other marine mammals in the Northern Hemisphere.

FIGURE 3.6.12

Mercury and cadmium levels in the livers of St. Lawrence belugas (1988-1990)

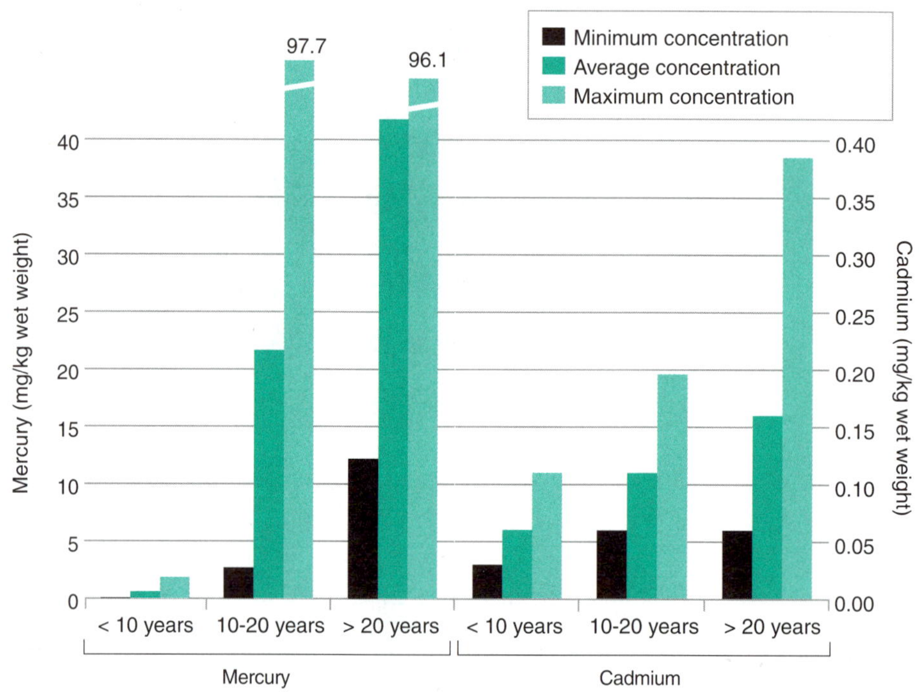

Note: Sample by age group. < 10 years: 5 belugas; 10 to 20 years: 13 belugas; > 20 years: 16 belugas.

Source: Based on data from Béland et al., 1992.

Of 34 belugas analysed between 1988 and 1990, five were under 10 years old, 13 were 10 to 20 years old and 16 were over 20 years old. Mercury levels in the belugas ranged from 0.07 mg/kg (wet weight) to 97.66 mg/kg. Average mercury levels ranged from 0.58 mg/kg for belugas aged 10 and under, to 21.64 mg/kg for belugas aged 10 to 20, and 41.76 mg/kg for belugas over 20. The low cadmium levels in these belugas (from 0.03 to 0.385 mg/kg wet weight) do not appear to be a cause for concern. Average cadmium levels ranged from 0.06 mg/kg (belugas aged 10 and under) to 0.11 mg/kg (belugas aged 10 to 20), to 0.16 mg/kg for belugas over age 20.

Information Supplement

PCB LEVELS IN THE EGGS OF DOUBLE-CRESTED CORMORANTS ON ÎLE AUX POMMES

Polychlorinated biphenyls (PCBs) were employed in a wide variety of industries before their use was restricted in 1972. Used in electrical equipment, printing inks, lubricants and insecticide sprays, PCBs are now found in every terrestrial and aquatic environment in the world, their worldwide dissemination ensured by atmospheric transport. A few of the molecule's 209 isomers are suspected to be highly toxic and cause metabolic disorders in a number of animal species, including aquatic birds. Because of their solubility in fatty tissues and great stability, PCBs tend to biomagnify in food webs. Since they are at the top of their food chain, fish-eating birds like the Northern gannet (*Sula bassana*) and Double-crested cormorant (*Phalacrocorax auritus*) are good indicators of trends in accumulation of these contaminants in the ecosystem.

Since 1972, the eggs of the Double-crested cormorant on Île aux Pommes, near Trois-Pistoles in the St. Lawrence Lower Estuary, have been sampled for analysis of PCB levels. Data in Figure 3.6.13 show that, after reaching a peak around 1976, PCB concentrations declined steadily to a level not currently considered problematic.

The decreasing trend in recent years is corroborated by relatively low PCB levels in Double-crested cormorant eggs harvested on other islands in the Lower Estuary in 1991 and 1992. Levels ranged from 1.32 to 5.01 mg/kg (mean 3.39±0.8) on islands in the Upper Estuary and from 1.10 to 4.01 mg/kg (mean 2.37±0.3) on islands in the Lower Estuary and Gulf.

FIGURE 3.6.13
PCB levels in the eggs of Double-crested cormorants on Île aux Pommes (1972-1992)

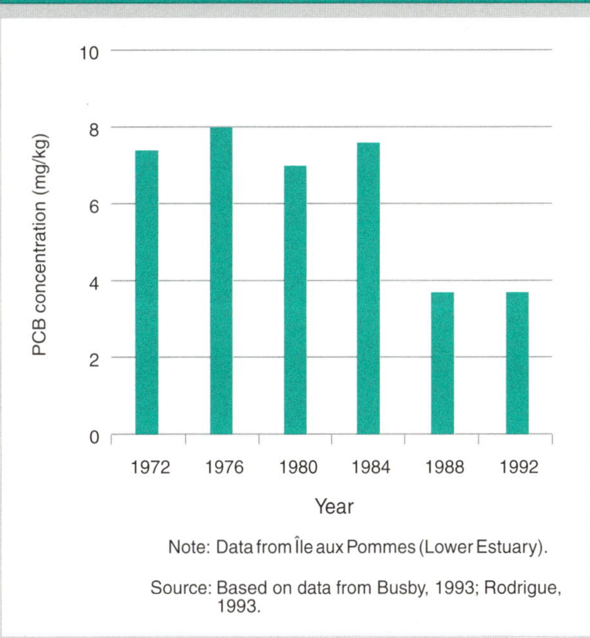

In the 1970s, eggs of Double-crested cormorants in the Great Lakes – the industrial heartland of the continent – had PCB levels as much as 5 to 10 times higher than those recorded in the Lower Estuary of the St. Lawrence in 1976, the worst year on record. Very high levels of PCBs, coinciding with equally high levels of DDE and other organochlorines, were then associated with a complete interruption in the reproduction of Double-crested cormorants. Such drastic effects were never seen in the Lower Estuary of the St. Lawrence.

Source: Based on data from Busby, 1993; Rodrigue, 1993.

CONDITION OF BIOLOGICAL RESOURCES

Main Findings

- The serious decline in abundance observed between 1986 and 1992 among some freshwater and saltwater fish stocks and species (some of which are at risk) occurred concurrently with relatively heavy commercial fishing, particularly of Rainbow smelt, American eel, Atlantic tomcod and Atlantic cod.
- Between 1969 and 1992, the Northern gannet population increased gradually, while the Greater snow goose population increased rapidly.
- The population of Beluga whales, an endangered species, remained steady at about 500 individuals between 1982 and 1992.
- A gradual decrease in contamination was observed for American eels caught at Kamouraska, Northern gannet eggs from Bonaventure Island and shrimp in the Saguenay Fjord, though contamination by toxic substances was still evident in the late 1980s in the nine species examined.

Relative Importance

■ This is the most-influenced characteristic among those selected for assessing the state of the St. Lawrence. It also ranks fifth among the eight most influential characteristics for the St. Lawrence as a whole. This means it is a hybrid characteristic.

Recommendations on Monitoring

- Separate indicators were used to measure abundance and contamination; new ways of integrating the measurement of these two aspects should be developed to take into account the diverse components of the St. Lawrence ecosystem.
- As scientific knowledge evolves, improved selection criteria should be developed so that the most representative species in different taxons can be selected.

3.7

SHIPPING

Canadian Coast Guard, Laurentian Region

Background

Navigation is especially difficult at certain points on the St. Lawrence, owing to areas of strong current, narrowing of the ship channel, and shallow water depth. The main environmental risks associated with shipping are spills of dangerous goods, the effects of lapping (increased wave action on the shoreline within one kilometre of a passing ship), and the introduction of exotic species during ship deballasting operations.

The introduction of exotic species was examined in the section on biodiversity, when the annual colonization of part of the River by Zebra mussels was discussed. The effects of lapping on the shoreline are considerable at specific points along the main course of the River. In terms of the River as a whole, however, the impact of this aspect of shipping is not significant. The topic is not discussed in this report, but is covered in Volume 1.

Historical data show that accidental spills occur when ships are refuelled in port, when vessels' double hulls are pumped out, when tankers are unloaded, or as a result of shipping accidents.

The risk of an accidental spill of dangerous goods is therefore the aspect that will be considered here, owing to its potentially disruptive effects, especially on the condition of biological resources, water quality, sediment quality and access to the shoreline and River. While it is not possible at present to assess clearly this environmental risk, three indicators provide information and represent the basic elements of this approach. The first indicator yields information on the percentage of dangerous goods handled in commercial ports in relation to total goods handled. The second tells us the proportion of marine trips made by merchant vessels and tankers whose draft exceeds the guaranteed water depth in two parts of the ship channel, and the third provides information on the number of spills recorded between 1991 and 1993, as well as on the average quantity spilled.

Interpretation

The most incident-prone areas, in view of the volume of dangerous goods, sailing difficulties and large amounts of oil being carried, lie along the two upstream stretches of the River. The Quebec City region is particularly sensitive to accidental spills, primarily owing to sailing difficulties (depth, tides, etc.) and the number of vessels that pass through it.

In 1992, ports on the Fluvial Estuary and Fluvial Section received close to 90% (more than 15 million tonnes) of the dangerous goods handled in the commercial ports of the St. Lawrence. Some 50% of all goods handled in the Fluvial Estuary are designated "dangerous goods". Furthermore, tankers' drafts most often exceed the guaranteed water depth in the North Traverse section of the ship channel near Île d'Orléans. Most of the spills reported between 1991 and 1993 involved oil.

Total tonnage and proportion of dangerous goods handled in commercial ports

Definition

Included in dangerous goods are all commodities handled bearing a UN identification number. UN numbers and their classification are taken from the United Nations' *Dangerous Goods Transportation Regulations*, endorsed by Canadian legislation (*Canada Gazette*, Chapter 34, August 21, 1992). This number represents a worldwide standard for characterizing and classifying dangerous goods, identifying them and handling them properly in all the world's ports.

According to UN classification, the following are considered dangerous goods:

Class 1: Explosives

Class 2: Gas

Class 3: Flammable liquids

Class 4: Flammable solids: substances which can spontaneously ignite or which, on contact with water, emit flammable gases

Class 5: Oxidizing agents and organic peroxides

Class 6: Poison and infectious substances

Class 7: Radioactive materials

Class 8: Corrosives

Class 9: Miscellaneous products or substances (includes miscellaneous dangerous goods, substances or products representing sufficient risk to justify regulating their transportation, but which cannot be assigned to any of the other classes). Substances representing an environmental risk and hazardous waste are also included.

Statistics on total tonnage by hydrographic region of the River are available from 1980 to 1992. Statistics concerning the tonnage of dangerous goods handled are available for 1992 only.

Limitations

Data on the volume of goods handled (including dangerous goods) in the commercial ports of the St. Lawrence should be complemented by information concerning the transshipment method used; volume alone is not sufficient to provide an accurate picture of the environmental risk involved.

Results

Problem Or Risk-Prone Sectors

The data in Figure 3.7.1 show that the largest volume of goods were handled in the ports of the Lower Estuary and Gulf. In 1992, 49 million tonnes of goods were handled there, or slightly more than the total volume handled in the other three sectors of the River combined. In the Fluvial Section, the volume handled in 1992 was about 22 million t, down some 20% since 1988. In the Fluvial Estuary, the volume was 19 million t, unchanged since 1984. The Upper Estuary and Saguenay were marginal sectors, with 5 million t of goods handled.

Most of the dangerous goods handled in 1992 were in the two upstream sectors, with close to 50% for the Fluvial Estuary and 28% for the Fluvial Section. In the Lower Estuary and Gulf, only about 5% of the commodities handled were classified as dangerous goods, although these were the sectors where the volume of goods handled was highest. The Upper Estuary and Saguenay accounted for a mere 2% of the tonnage of dangerous goods handled.

FIGURE 3.7.1
Total tonnage and proportion of dangerous goods handled in certain commercial ports of the St. Lawrence (1980-1992)

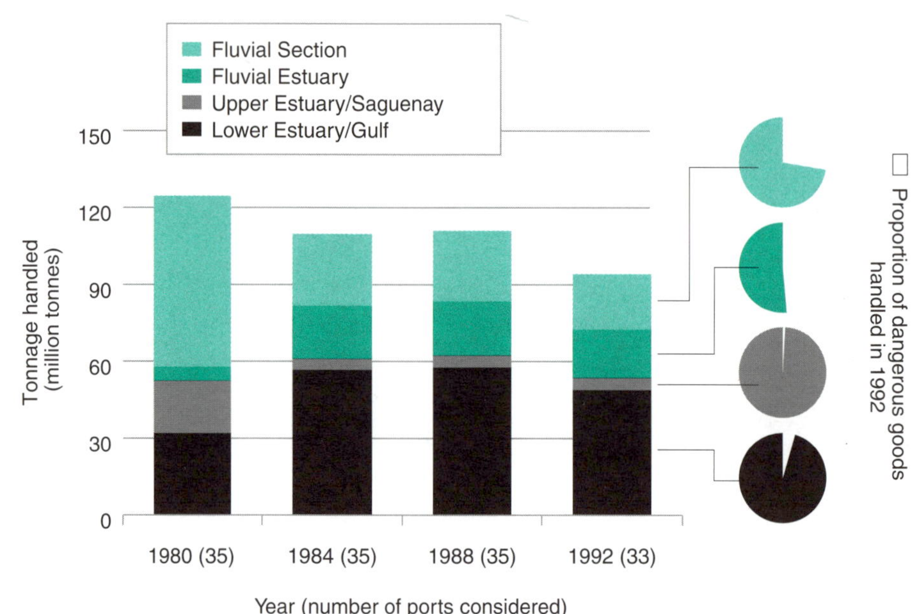

Source: Based on data from Canadian Coast Guard, 1993.

Proportion of trips by merchant vessels and tankers exceeding guaranteed water depth in two parts of the ship channel

Definition

For 1988-1992, the data gathered provide information on the percentage of trips made by vessels whose draft exceeded the guaranteed water depth. Two types of ships are considered: merchant vessels and tankers. The sectors of the ship channel covered are the North Traverse (Île d'Orléans) and the Montreal–Quebec City stretch.

Guaranteed water depth

Guaranteed water depth means the shallowest depth of the ship channel corresponding to the water depth at low tide, i.e., when the water level is equal to zero on the chart level. This water depth is ensured at all times by dredging.

However, the actual water depth is usually greater than this, because of the tide.

Merchant vessels

All vessels carrying dry goods or packaged liquid merchandise.

Tankers

All vessels carrying bulk liquid goods (not packaged in barrels or other types of containers), regardless of the substance.

Limitations

This measurement reflects a routine practice whereby the tides are used to sail through certain sectors.

Problem or Risk-Prone Sectors

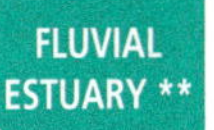

*Lake Saint-Louis, lesser La Prairie basin, and upstream sector of Lake Saint-Pierre.

**Region between Quebec City and Île aux Oies.

Results

A review of the data on marine trips made by vessels whose draft exceeds the guaranteed water depth for 1988-1992 shows that the number of such trips has fallen by 20% since 1988. This drop tallies with the fall in the total number of marine trips, which went from 12 434 in 1988 to 10 461 in 1991. Over that period, an average of 2.6 times as many trips exceeding the guaranteed water depth were recorded in the North Traverse (Île d'Orléans) sector than along the Montreal–Quebec City stretch.

Taking both sectors together, more tankers made trips while exceeding the guaranteed water depth, except in 1991. In the North Traverse sector, an average of 70% of trips exceeding the guaranteed water depth were made by tankers, and 30% by merchant vessels. In the Montreal–Quebec City sector, the opposite was true: on average, 82% of trips exceeding the guaranteed water depth were made by merchant vessels, and 18% by tankers.

Montreal–Quebec City sector: The total number of trips by merchant vessels in the Montreal–Quebec City sector fell by 19% between 1988 and 1992, from 5066 to 4100. Tankers declined by 17% between 1988 and 1992, from 1365 to 1131. The percentage of trips by merchant vessels exceeding the guaranteed water depth fell overall, from 0.69% in 1988 to 0.22% in 1992, except in 1991, when it increased (0.67%). The same pattern was seen with tankers, for which the percentage exceeding the guaranteed water depth fell continually, from 0.66% in 1988 to 0.09% in 1992, with the exception of 1991, when it increased to 0.56%.

North Traverse sector: The total number of trips by merchant vessels sailing through the North Traverse fell by 13% between 1988 and 1992, from 4769 in 1988 to 4141 in 1992. The total number of tankers transiting this sector fell by 29%, from 1234 in 1988 to 880 in 1992. The percentage of trips by merchant vessels exceeding the guaranteed water depth rose, however, from 0.36% in 1988 to 0.60% in 1992. The same pattern was seen with tankers, for which the percentage exceeding the guaranteed water depth has risen constantly since 1988 (except in 1991, when it dropped), climbing from 3.48% in 1988 to 5.45% in 1992.

FIGURE 3.7.2

Annual number of marine trips by vessels with a draft exceeding the guaranteed water depth (1988-1992)

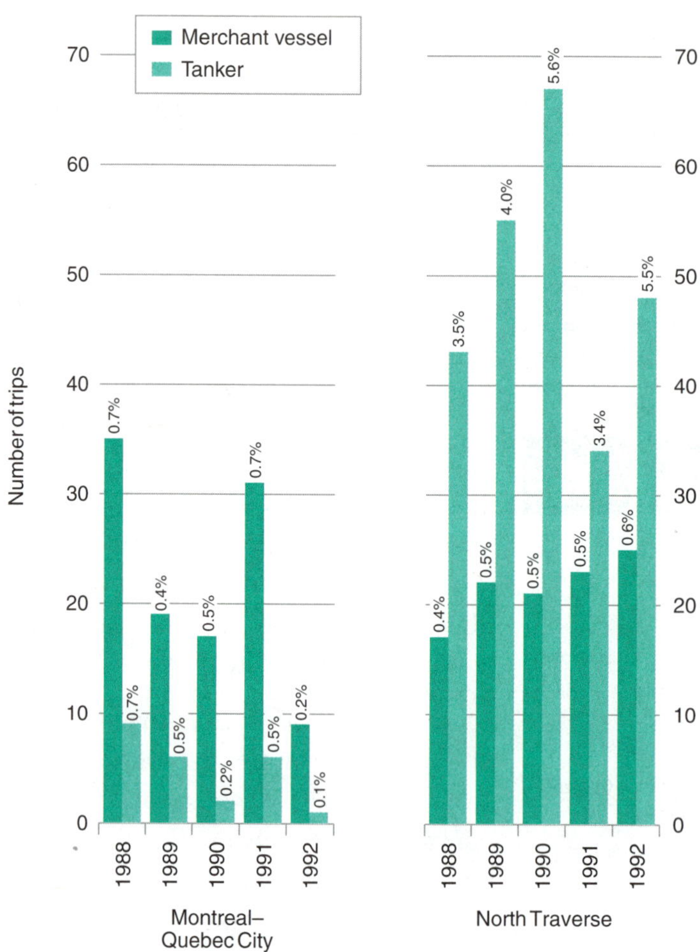

Note: Figures above bars represent trips exceeding guaranteed water depth as a percentage of the total number of trips.

Guaranteed water depth for Montreal–Quebec City: 10.7 metres until 1991 and 11.0 metres in 1992; guaranteed water depth for North Traverse: 12.5 metres.

Source: Based on data from Canadian Coast Guard, 1993.

Number of accidental spills recorded

Definition

Recorded accidental spill

Any pollution-generating (oil or other hazardous chemicals) spill reported to Vessel Traffic Services by the Canadian Coast Guard Alert Network that is the subject of a pollution incident report, whether of marine (ship), land-based or unknown origin.

Recorded spills occurred in navigable waters managed by the Canadian Coast Guard, Laurentian Region.

Limitations

The data needed to monitor spills from vessels plying the St. Lawrence are limited. They come from the Alert Network, which records pollution-generating spills in the Laurentian Region, from Cornwall to Anticosti Island.

The number of spills is not monitored on a regular basis. Only spills large enough to trigger a pre-alert are recorded. Average tonnage and spill source are the only information provided on the nature and impact of these spills, which can occur during transshipment of oil or chemical products or during refuelling operations. In addition, illegal discharges may take place during tank cleaning operations (deballasting), for instance.

The number of pollution-generating spills reported each year may vary according to public awareness of the problem. Today, the public are more likely to report spills than they used to be.

Problem or Risk-Prone Sectors

FLUVIAL SECTION | FLUVIAL ESTUARY

UPPER ESTUARY AND SAGUENAY | LOWER ESTUARY AND GULF

Results

The compilation of recorded spill (or pollution incident) reports shows that between 1991 and 1993, the number of pollution incidents involving vessels ranged from 61 to 97, and those associated with an unknown or land-based source ranged from 34 to 83 (see Figure 3.7.3). No trend can be isolated over a three-year period, but a substantial increase in the number of incidents reported in 1992 should be noted.

Most reports recorded involved oil spills. The average quantity of oil spilled by vessels rose from 255 gallons in 1991 to 340 gallons in 1992 and 475 gallons in 1993, for an 86% increase in three years.

FIGURE 3.7.3

Number of accidental spills recorded on the St. Lawrence, from Cornwall to Anticosti Island, and average quantity spilled (1991-1993)

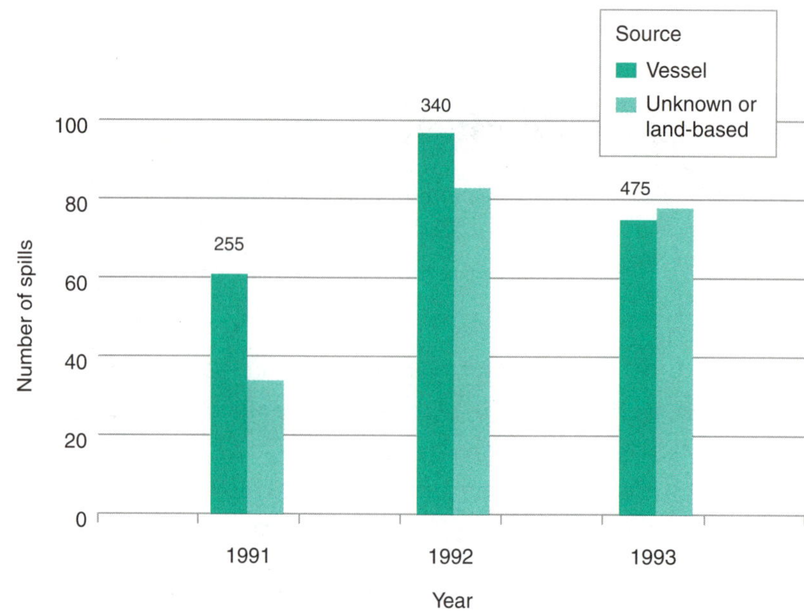

Note: Figures above bars indicate average quantity spilled (in gallons).

Source: Based on data from Canadian Coast Guard, 1994.

SHIPPING

Main Findings

- In 1992, 17 million tonnes of dangerous goods (out of a total tonnage of some 95 million t) were handled in the River's commercial ports; close to 90% of the tonnage was handled in the Fluvial Estuary and Fluvial Section.
- In 1992, 48 marine trips (out of a total of 880) made by tankers in the North Traverse of the ship channel involved a draft in excess of the guaranteed water depth. Since 1988, the proportion of trips exceeding this water depth has fluctuated between 3.36% and 5.57%.
- In 1993, 75 accidental spills were recorded from vessels and 78 from land-based or unknown sources. The average quantity of oil spilled by vessels was 475 gallons.

Relative Importance

- Shipping is not one of the eight most influential characteristics for the St. Lawrence as a whole, since its influence is felt primarily in the main course of the River and Fluvial Estuary.

Recommendation on Monitoring

- The measurements selected for monitoring changes in this characteristic help to assess the environmental risk associated with shipping. However, a more comprehensive index should be developed, that takes into account, for instance, the nature of the product spilled, the characteristics of the waterway (identification of sectors presenting risks for certain types of vessels), and the clear link between vessel age and the probability of serious damage leading to a wreck and disaster.

3.8

MODIFICATION OF THE FLOOR AND OF HYDRODYNAMICS

Environment Canada, R. Rochon

Background

Past experience has shown the possible environmental impacts of major dredging operations and changes in hydrodynamics (see information supplement entitled, "Environmental consequences of the extensive dredging work carried out between 1952 and 1970"). Among other things, dredging releases toxic substances when it is carried out in contaminated sediment, resuspends solids which alter the characteristics of the water and causes changes to the flow regime. The first of these effects has an impact on the overall ecosystem, while it is suspected that the other two act directly or indirectly on fish by disturbing migration patterns of anadromous species or physically altering breeding and feeding habitats.

Today, the dredging activities necessary to maintaining port facilities and the ship channel can contribute to resuspending contaminated sediment in open water. The sediment then acts as a carrier for pollutants which will affect water quality and the condition of biological resources.

Interpretation

Between 1983 and 1991, the average annual volume dredged fell by 300%. Some 600 000 m^3 of sediment was moved each year to maintain commercial shipping in the St. Lawrence, equivalent to 60 000 trips by trucks with a capacity of 10 m^3 each. The largest amounts were dredged in the North Traverse, the Lower Estuary and the Gulf.

Average annual volume of material dredged

Definition

The only measurement currently available for conducting a partial assessment of this characteristic is the average annual volume of sediment dredged.

The data are available by sector, from 1983 to 1991. This measurement offers general information on the hydrographic regions where there are changes in the floor and hydrodynamics of the River.

Limitations

The indicator used does not provide any information on the sites, the particle-size distribution of the material or the degree of contamination. It does not take other activities, such as filling, into account.

Problem Or Risk–Prone Sectors

FLUVIAL ESTUARY*

LOWER ESTUARY AND GULF

*Particulary North Traverse sector.

Results

The annual volume of material dredged stands between 400 000 and 600 000 m³ for the St. Lawrence as a whole (see Figure 3.8.1). In 1983 and 1988, however, the volume was much greater, with close to 1.3 million and 900 000 m³ dredged, respectively. The Lower Estuary and Gulf consistently account for the largest dredging volumes, closely followed by the Fluvial Estuary. In these two sections of the River, the annual volume dredged is from 200 000 to 300 000 m³. The Fluvial Section and Upper Estuary post the lowest volumes, generally dredging less than 100 000 m³ per year. Between 1983 and 1991, 812 520 m³ was dredged in the North Traverse of the ship channel alone, near Île d'Orléans, for an average of 90 280 m³ per year.

FIGURE 3.8.1

Average annual volume dredged (m³) in the St. Lawrence (1983-1991)

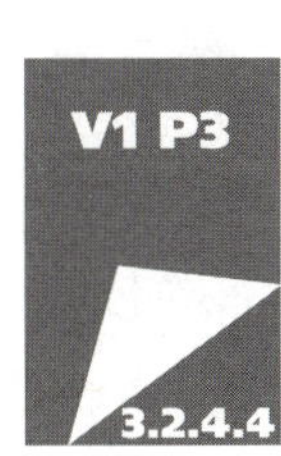

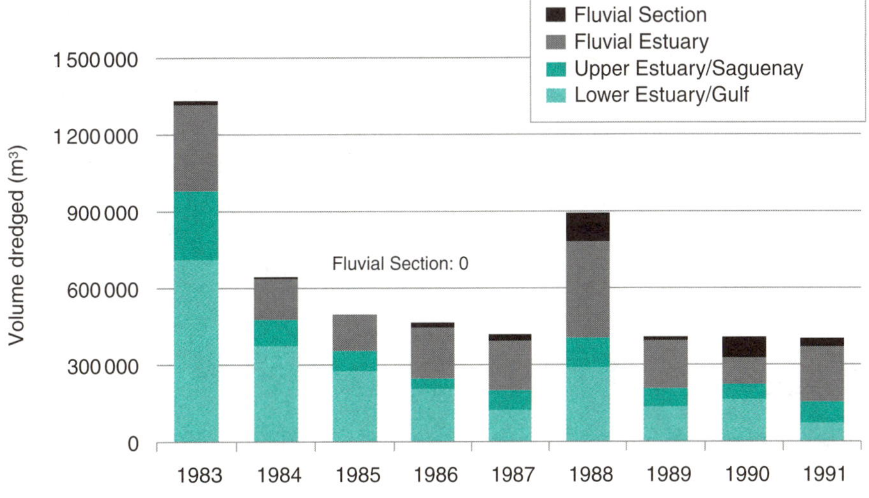

Source: Based on data from Olivier and Bérubé, 1993.

Information Supplement

ENVIRONMENTAL CONSEQUENCES OF THE EXTENSIVE DREDGING WORK CARRIED OUT BETWEEN 1952 AND 1970

The environmental consequences of dredging sediment to maintain shipping activities were not really a concern of St. Lawrence managers 25 years ago. There was barely a suspicion that the extensive dredging work associated with widening the ship channel (between 1952 and 1970) and construction of the St. Lawrence Seaway (between 1954 and 1959) might have played a direct or indirect role in the precarious situation of certain fish stocks. The widening and digging of the ship channel between Cap Tourmente and Montreal meant the blasting and removal of millions of cubic metres of sediment over a 200-km stretch.

With hindsight, researchers consider these major operations and the changes they wrought to the hydraulic regime to have either caused the direct mortality of a large number of fish (Robitaille et al. 1988), or to have had an indirect but equally devastating environmental impact. Few researchers doubt the role that changes of this magnitude to the river floor have had in the reduction of stocks or the critical situation of the American shad and Striped bass, as well as in the decline of the Lake sturgeon and Atlantic sturgeon.

Today, dredging activities are subject to other controls, and practices are changing. For example, operations are authorized only on certain dates, prior impact assessments are conducted, and new dredging and sediment deposition methods have been implemented.

Source: Robitaille et al., 1988; Marquis et al., 1991.

MODIFICATION OF THE FLOOR AND OF HYDRODYNAMICS

Main Findings

- Dredging to maintain ports, the ship channel and the St. Lawrence Seaway alters the bed and flow of the River; between 400 000 and 600 000 m^3 of sediment has been dredged yearly since 1989. The largest volume is removed from the Fluvial Estuary and the Lower Estuary. Little is known about the impact of this activity on habitats, water quality and the condition of biological resources. Among other things, the work is likely to resuspend contaminated sediment, which can be absorbed by organisms, and to bring about physical changes in habitats.
- The construction of riverside infrastructure providing access to the River for recreational or port activities contributes to the physical alteration of habitats. This is also true for the fill and earth removal activities associated with construction of riverside highway infrastructure and urban growth.

Relative Importance

- Modification of the floor and hydrodynamics do not emerge as one of the most influential characteristics for the River as a whole, but they do have a marked impact in the Fluvial Section and Fluvial Estuary.

Recommendation on Monitoring

- An indicator remains to be developed for weighting dredged volume. The indicator would incorporate information on such factors as the site, the area affected, particle-size distribution and toxic load (and toxic substance content) of the material removed, as well as a risk assessment for biological resources downstream or nearby, and habitat changes. It should also take into account any fill activities (for various purposes) which could also alter the bed and hydrodynamics of the River.

3.9

SHORELINE MODIFICATIONS

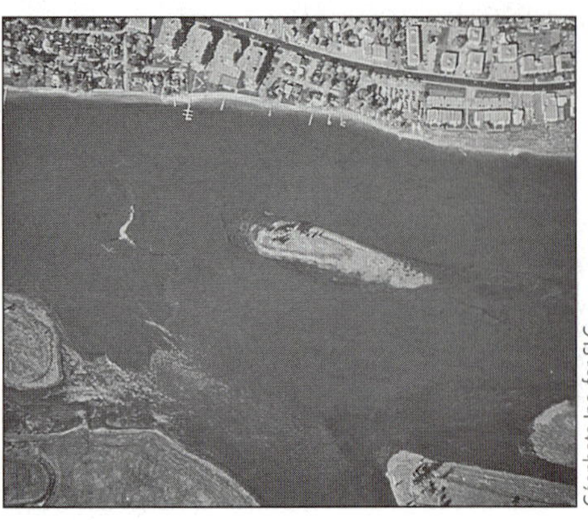

Géophoto Inc. for SLC

Background

The role of wetlands in biological productivity, their importance in certain critical phases of the life cycle of several species of fish, amphibians and waterfowl and the ecosystem's capacity for self-purification are well known. Given their ecological and economic importance, changes in these areas (caused primarily by encroachment, especially on the outskirts of the island of Montreal) have a strong impact on the state of the fluvial ecosystem. The effects of certain structures (such as the Portneuf and Bécancour wharves) on migratory fish following water flow changes have been documented. Encroachment on wetlands along the St. Lawrence peaked between 1960 and 1980 (see information supplement entitled, "Wetland losses along the St. Lawrence from 1945 to 1976").

Interpretation

According to Gratton and Dubreuil (1990), the distribution of wetlands along the St. Lawrence between 1980 and 1986 shows that, of the 79 700 hectares of wetlands, 79% (63 000 ha) are located in the Fluvial Section, 5% along the Fluvial Estuary, 8% along the Upper Estuary and 8% along the Lower Estuary and Gulf. In the Fluvial Section, the Lake Saint-Pierre region comprises 32 280 ha, or 41% of all St. Lawrence wetlands. The Lake Saint-François region follows, with 11 300 ha (14%), while the wetlands of the estuary and Gulf occupy 16 694 ha (21%). In 1986, no information was available on the area of wetlands along the Saguenay River.

Over the past three decades, wetlands have declined substantially in area, particularly along the Fluvial Section and Fluvial Estuary. There are no data available to show whether other losses or gains have occurred since the late 1970s. The current spatial distribution of wetland areas shows that the two upstream sectors are especially vulnerable to changes, in particular the Fluvial Section and its richest stretch, the Lake Saint-Pierre area.

Surface area of wetlands, in hectares

Definition

The measurement, in hectares, of the area occupied by wetlands was selected as an indicator of the state of the environment. It is used to determine the spatial significance of wetland areas along the River. Identifying how the measurement changes over time will help evaluate losses or gains in surface area.

Examined in conjunction with data on the condition and biodiversity of biological resources, and data on access to the shoreline and River, this indicator helps provide a more accurate picture of shoreline changes and the fluvial ecosystem's condition.

Limitations

Overall encroachment on wetlands is known only for the period 1945-1976. Lack of data makes it impossible to gauge whether further losses have occurred since 1976. This indicator is therefore a reference point which will, for a subsequent assessment, serve to show what further shoreline changes have taken place. The data presented here were obtained in 1990 and 1991, and cover only the wetlands in the River corridor (Cornwall to Montmagny sector).

PROBLEM OR RISK-PRONE SECTORS

* No inventory.

Results

In 1990, the Fluvial Section totalled some 48 000 ha of wetlands along a strip up to 1-km-wide on either side of the St. Lawrence. One-half of this area consisted of plant communities, or aquatic grass beds, while the remainder was divided into three other groups: marsh, wet meadow and swamp. A significant difference in values is observed between the Fluvial Section and the Fluvial Estuary. In the latter, wetlands cover an area of some 9500 ha, consisting mainly of marshland. Swamp and wet meadow account for less than 1700 ha.

FIGURE 3.9.1

Surface area of St. Lawrence wetlands (1990-1991)

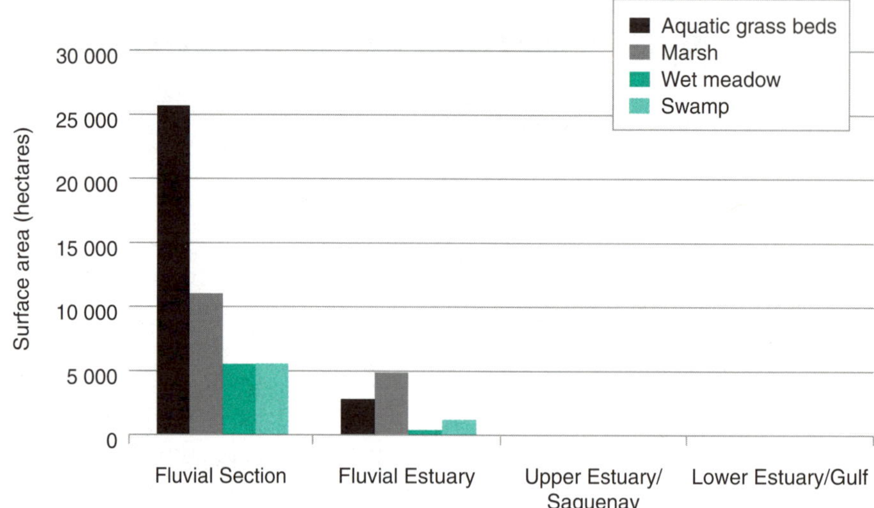

Note: Where no bar appears, the data were unavailable.

Sources: Based on wetland maps produced using MEIS-II airborne remote sensing images by Aménatech, 1991, 1992a, 1992b.

Information Supplement

WETLAND LOSSES ALONG THE ST. LAWRENCE FROM 1945 TO 1976

Between 1945 and 1976, a total of 3649 ha of wetlands were lost, 75% of them between 1945 and 1960.

A little over one-third of these wetland losses can be attributed to the reclamation of arable land from flood zones (1231 ha) and, to a lesser extent, from fill activities (970 ha). Some 1448 ha of wetlands were lost to residential construction (431 ha), earth removal for the construction of locks and completion of the St. Lawrence Seaway (346 ha), construction of service infrastructures (337 ha), industrial and commercial construction (285 ha) and leisure infrastructures (49 ha).

FIGURE 3.9.2
Wetland losses along the St. Lawrence, by type of change (1945-1976)

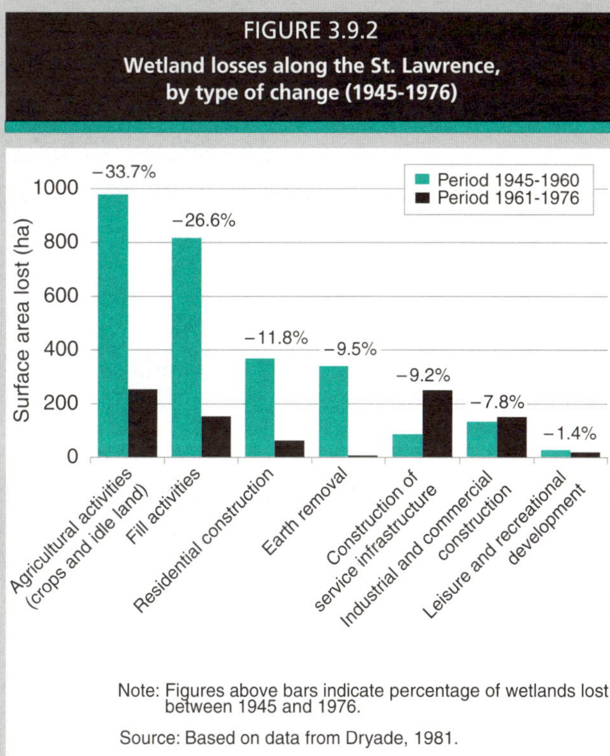

Note: Figures above bars indicate percentage of wetlands lost between 1945 and 1976.

Source: Based on data from Dryade, 1981.

The concentration of wetlands, combined with that of the population and economic activity in the same sections, are exerting strong pressure on wetlands in the River corridor, particularly along the Fluvial Section and the Fluvial Estuary. Between 1960 and 1975, a number of structures encroached on the natural banks of the St. Lawrence over a cumulative length of 175 km between Montreal and Quebec City.

Since that time, the threat to wetland habitats has been considerably alleviated with the adoption of legislation, new regulations and greater public awareness. But encroachment on wetlands by agriculture, residential and port construction and industry continues. For instance, the construction of a highway on the Beauport flats and the invasion of the Kamouraska swamp for agricultural purposes have altered the St. Lawrence shoreline within the past 15 years.

Source: Based on data from Dryade, 1981.

SHORELINE MODIFICATIONS

Main Findings

- The St. Lawrence has undergone substantial wetland losses, amounting to 3649 ha, between 1945 and 1976. It is impossible to assess the surface area lost or gained since 1976.
- The only inventory available (1990-1991) indicates that wetlands total 48 000 ha along the main course of the River and Fluvial Estuary.
- In view of past losses and the acknowledged ecological and economic significance of wetlands, additional surface area losses could have a major impact on the condition of the fluvial ecosystem, including loss of essential riparian wildlife habitats, disappearance of plant species, reduction of the system's capacity for self-purification, and loss of biodiversity.

Relative Importance

▲ The shoreline modifications indicator is ranked last among the eight most influential characteristics.

Recommendation on Monitoring

- Over the coming years, the surface area of wetlands will have to be closely monitored. Weighting by means of biological characteristics (e.g. richness of flora and fauna, biological productivity levels and uniqueness) would give more weight to these assessments and help provide a clearer understanding of the impact on the River of stresses caused by the loss of wetlands.

3.10

URBAN WASTEWATER DISCHARGES

Parks Canada, J. Audet

Background

Urban wastewater discharge is usually associated with bacterial contamination of the water and risks for the health of the communities living downstream of the outfall. This wastewater has a direct effect on the quality of water for human consumption, contact recreational activities (such as swimming, windsurfing and fishing) and the salubrity of shellfish-producing waters.

Some 15 years ago, the Comité d'étude sur le fleuve Saint-Laurent profiled the bacterial contamination of the River, which was already substantial in the mid-1970s and increased up until the early 1980s. The gradual introduction of a water treatment infrastructure made it possible to treat the urban wastewater of a growing proportion of the riverside population.

Interpretation

The proportion of riverside municipalities treating their wastewater has been on the rise since 1986. In 1992, it stood at more than 30%, accounting for 65% of the population served by treatment plants.

The proportion of riverside municipalities treating their wastewater and the percentage of the riverside population served

Definition

By counting the number of riverside municipalities treating their wastewater and evaluating the riverside population served by treatment plants, we can estimate how this characteristic has evolved. The available data cover the 1986-1992 period.

Limitations

The proportion of the riverside population served by treatment plants was slightly underevaluated, since it was calculated in relation to the overall riverside population and not to the riverside population connected to a sewer system. In 1991, an estimated 95% of the total population between Valleyfield and Quebec City were connected to a sewer system.

The two measurements selected for this indicator allow for only a partial assessment, since they provide no information on the volume discharged or on the discharge's nutrient, contaminant and coliform load.

Nor is the treatment of urban wastewater a guarantee against contamination. While, as a general rule, treatment substantially reduces or eliminates inputs of coliforms into the River, some treated wastewater outfalls (wastewater outfall from an aerated lagoon or activated-sludge treatment plant, for instance) may carry a considerable load of dissolved organic matter, bacteria and toxic substances (lead, copper, zinc, hydrocarbons, surfactants and solvents).

Results

PROBLEM OR RISK-PRONE SECTOR

In 1992, the riverside population served by treatment plants was distributed as follows: 73% along the Fluvial Section, 19% in the Fluvial Estuary, 5% in the Upper Estuary and Saguenay, and 3% in the Lower Estuary and Gulf. In 1992, an estimated 1.35 million people living along the River continued to discharge their wastewater into the River without any treatment. This population broke down as follows: 72% along the Fluvial Section, 5% in the Fluvial Estuary, 9% in the Upper Estuary and Saguenay, and 14% in the Lower Estuary and Gulf.

In 1986, only 10% of the riverside population (16 municipalities), or approximately 418 000 people, were served by treatment plants (see Figure 3.10.1). Between 1986 and 1989, the population served rose to 1 322 000 (52 municipalities), a 23% increase in three years. Between 1989 and 1992, the population served rose to 2 615 000 (107 municipalities), up 32% in three years, thus reaching 65% of the overall population of the 339 riverside municipalities.

Between 1989 and 1992, 55 riverside municipalities, among them some of the most densely populated in Quebec, joined those already treating their wastewater. They include the municipalities of the Quebec Urban Community, Lévis–Lauzon, the municipalities of the Trois-Rivières and Sorel region, and a large number of Montreal south shore municipalities, including Longueuil, Boucherville, Brossard, Saint-Lambert and Châteauguay.

FIGURE 3.10.1

Proportion of riverside municipalities treating their wastewater and proportion of the riverside population served by treatment plants (1986-1992)

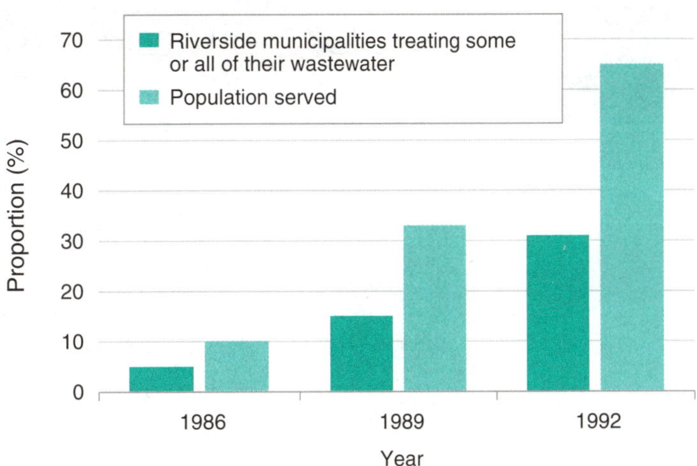

Note: Percentages calculated for 339 riverside municipalities and a total population of 4 044 000 in all cases.

Source: Based on data from MENVIQ, 1992b.

URBAN WASTEWATER DISCHARGES

Main Findings

- Between 1986 and 1992, the proportion of the riverside population served by wastewater treatment plants rose from 10% to 65%. However, in 1992, 1.35 million inhabitants, or 33% of the riverside population, continued to discharge their untreated wastewater directly into the River. Treated urban wastewater may also contribute to contamination by carrying a load of dissolved organic matter and bacteria via outfalls.
- At present, we cannot gauge the contribution of treated or untreated urban wastewater to the toxic load in the River. These discharges include oils and greases, solvents, household products and waste from small industry.
- Urban wastewater discharges and agricultural activities are the main sources of bacterial contamination. The measurements available to us can not assess their relative importance.

Relative Importance

▲ For the St. Lawrence as a whole, urban wastewater discharge is ranked sixth among the eight most influential characteristics.

Recommendation on Monitoring

- This indicator could be enhanced by combining data on total volume and quality of treated discharges. The latter element is significant owing to the quantity of nutrients, contaminants associated with suspended solids and microbiological load which return in part to the River, depending on the wastewater treatment method and the effectiveness of treatment facilities. Note that the contribution of urban wastewater to the River's toxic load (oils and greases, solvents, household products and waste from small industry, legally or illegally connected to the system) cannot be assessed at present, though such an assessment would be desirable.

3.11

INDUSTRIAL WASTEWATER DISCHARGES

St. Lawrence Centre, S. Lorrain

Background

Fifteen years ago, researchers and managers had no accurate measurement tools for assessing toxic inputs; assessments were perfunctory and descriptive. There was also a tendency to believe that contaminants came from upriver – that is, from the Great Lakes, which had seriously deteriorated at that time. However, recent findings show that a significant contribution to toxic load comes from riparian activities downstream of Cornwall.

In the late 1980s, the Chimiotox Index and the Potential Ecotoxic Effects Probe (PEEP) were developed to assess toxic inputs and the toxicity of liquid effluent from 49 of the 50 plants targeted under the St. Lawrence Action Plan (one of the metallurgical plants closed down before its effluent could be characterized). Chimiotox and PEEP were selected for this report as indicators of the state of industrial wastewater discharges.

Interpretation

In 1992, 1 421 218 m^3 of process water was released each day by the 50 targeted plants: 56% came from pulp and paper mills and 35% from metallurgical plants. Chemical plants released 13.9 t/day of suspended solids into the St. Lawrence, pulp and paper mills 62.9 t/day, and metallurgical plants 195.3 t/day.

Toxic inputs from plants targeted in 1988 under the St. Lawrence Action Plan fell from 5.2 million units in 1988 to 1.3 million units in 1993. For all four sectors of the River put together, these inputs decreased more markedly for the metallurgy sector (down 87%) and the organic chemicals sector (down 86%). The most significant reductions were posted by the metallurgy sector in the estuary and Gulf, and by the organic chemicals sector in the Fluvial Section. Since 1988, there has been a substantial drop in inputs of heavy metals and non-chlorinated phenols. According to the Chimiotox index in 1982, PAHs and phthalates appear to have been removed from industrial discharges (see information supplement entitled, “Toxic substances released by industry”).

The Fluvial Section receives most of the toxic inputs from organic chemical, inorganic chemical and metallurgical plants. Toxic inputs from pulp and paper mills are found primarily in the Fluvial Estuary and the Upper Estuary and Saguenay.

According to the PEEP index, effluent from pulp and paper mills and metallurgical plants present the highest toxicity values for the River as a whole.

Chimiotox Index

Definition

The indicator selected, the Chimiotox index, is used to weight the importance of the liquid waste according to the relative toxicity of its content. By using standardized units, it is also possible to place a figure on, and to visualize, contamination patterns by industry, industrial sector, river sector, and year.

The evolution of the Chimiotox index over time is presented for all plants in the same industrial sector for 1988-1992. The industrial sectors represented are pulp and paper, metallurgy, organic chemicals and inorganic chemicals. The data are presented for each of the four hydrographic regions of the St. Lawrence.

Limitations

The Chimiotox index provides only a partial assessment of industrial wastewater discharges. It does not consider the capacity of the receiving environment, acidity levels or the synergistic or antagonistic effects of substances in interaction, or even conventional parameters such as suspended solids load, biological oxygen demand or chemical oxygen demand. Some 6300 manufacturing establishments whose industrial facilities are smaller than those of the 50 targeted plants stand along the shoreline of the St. Lawrence. The data available cover only 49 of the 50 plants considered a priority and targeted under the St. Lawrence Action Plan. In 1993, 43 of the 50 plants were still operating, six along the Fluvial Section (one in metallurgy, and five in chemicals); one mill along the Fluvial Estuary (pulp and paper) had closed down.

Results

Problem Or Risk-Prone Sectors

FLUVIAL SECTION | FLUVIAL ESTUARY | UPPER ESTUARY AND SAGUENAY | LOWER ESTUARY AND GULF

The sectors with the largest inputs are: the main course of the River and Fluvial Estuary, the upstream parts of the Upper Estuary and Saguenay, and the Baie-Comeau sector in the Lower Estuary.

The four industrial sectors are all represented in two of the River's hydrographic regions: the main course of the River and the Fluvial Estuary.

The main course of the River or Fluvial Section had the highest total toxic input (sum of toxic inputs from the four industrial sectors) between 1988 and 1993. In 1988, this input was estimated at 3.3 million Chimiotox units, or 64% of total toxic inputs for the four sectors of the River. In 1993, it reached 603 462 Chimiotox units and accounted for 45% of total toxic inputs for the entire River (see Figure 3.11.1). An 82% decrease in total toxic inputs from this sector of the River was noted between 1988 and 1993. The industrial sectors contributing the most to this reduction were metallurgy and organic chemicals. In 1993, 54% of liquid toxic waste in the Fluvial Section came from metallurgical plants (327 854 units), 39% from inorganic chemical industries (233 928 units), 7% from organic chemical plants and less than 1% from pulp and paper mills.

The total toxic input for the Fluvial Estuary was estimated at 875 830 Chimiotox units in 1988 and 326 108 units in 1993, or 17% and 25%, respectively, of overall toxic inputs for all four sectors of the River. Thus, a 63% decrease was seen in total toxic inputs for this sector of the River between 1988 and 1993. The metallurgy and pulp and paper sectors contributed the most to this reduction. However, in 1993, 96% of the liquid toxic waste in the Fluvial Estuary still came from pulp and paper mills (313 431 units), 3% came from organic chemical plants (9497 units), while the remainder came from inorganic chemical and metallurgical plants.

The total toxic input for the Upper Estuary and Saguenay was estimated at 503 378 Chimiotox units in 1988, and 350 511 units in 1993, or 10% and 26%, respectively, of total toxic inputs for the River as a whole. The total toxic input for this sector of the River therefore fell by 29% between 1988 and 1993. The industrial sector contributing the most to this reduction was metallurgy. In 1993, 85% of liquid toxic waste in the Upper Estuary and Saguenay came from pulp and paper mills (298 703 units), 14% from metallurgical plants (49 888 units) and approximately 1% from inorganic chemical plants (1920 units).

For the Lower Estuary and Gulf, total toxic input was estimated at 491 412 Chimiotox units in 1988 and 53 362 units in 1993, or 9% and 4% respectively, of the overall toxic inputs for the four river sectors. An 89% decrease in overall toxic input for this part of the River was thus seen between 1988 and 1993. Metallurgy was the industrial sector contributing the most to the reduction. In 1993, 97% (51 725 units) of the liquid toxic waste in the Lower Estuary and Gulf came from pulp and paper mills, with the remaining 3% (1637 units) coming from metallurgical plants.

FIGURE 3.11.1A

Chimiotox indexes by hydrographic region and industrial sector (1988-1993)

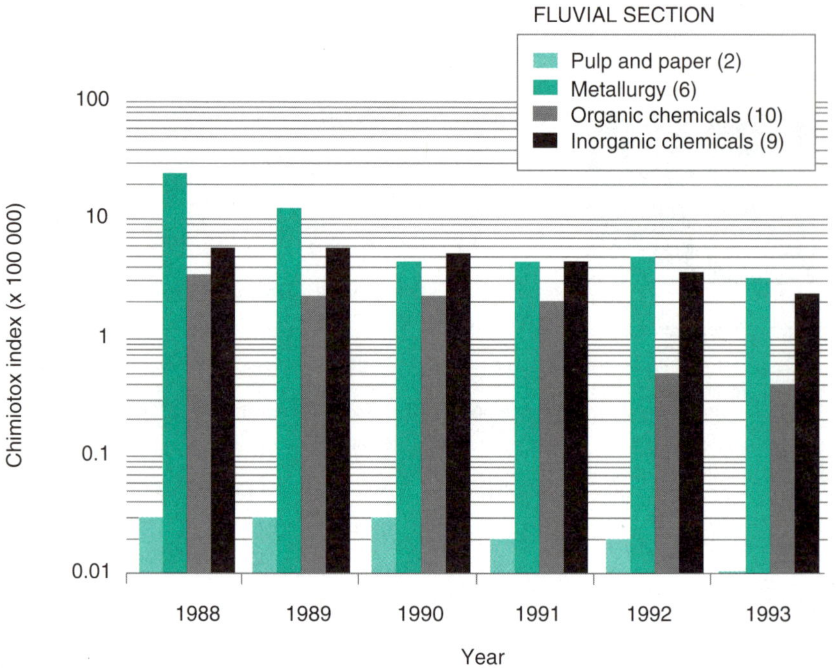

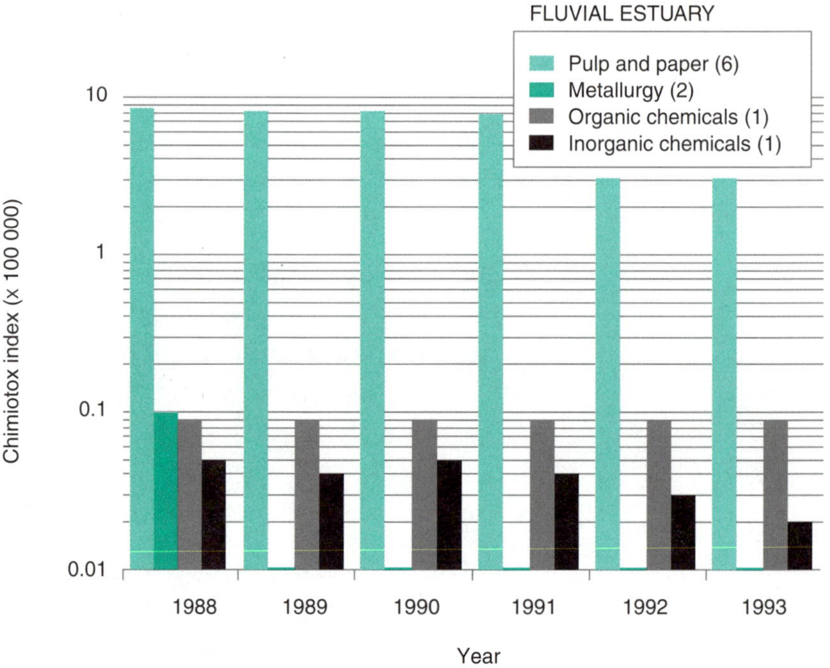

Note: Figures in brackets indicate the number of plants per industrial sector.

Source: Villeneuve,1994.

FIGURE 3.11.1B

Chimiotox indexes by hydrographic region and industrial sector (1988-1993)

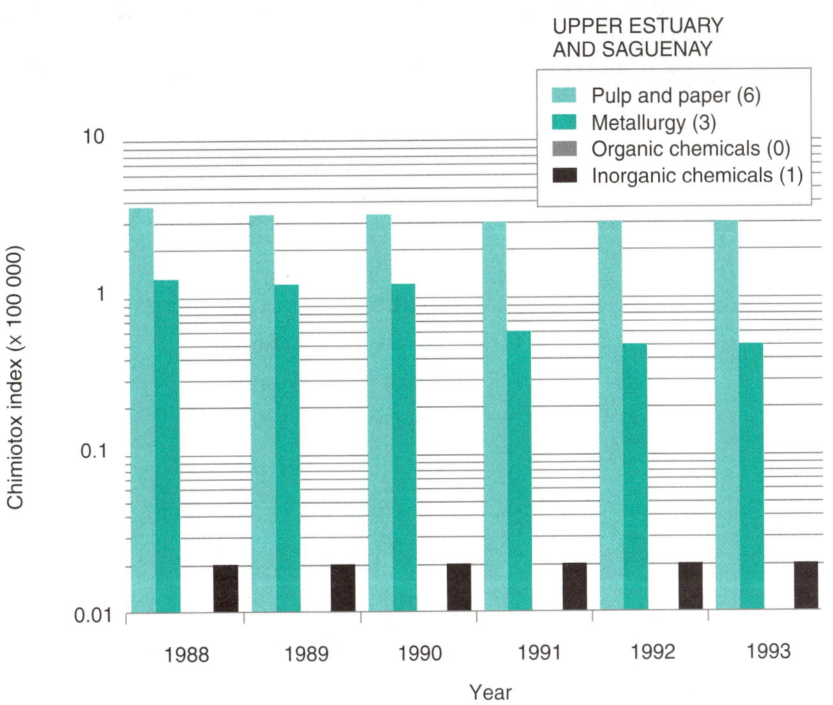

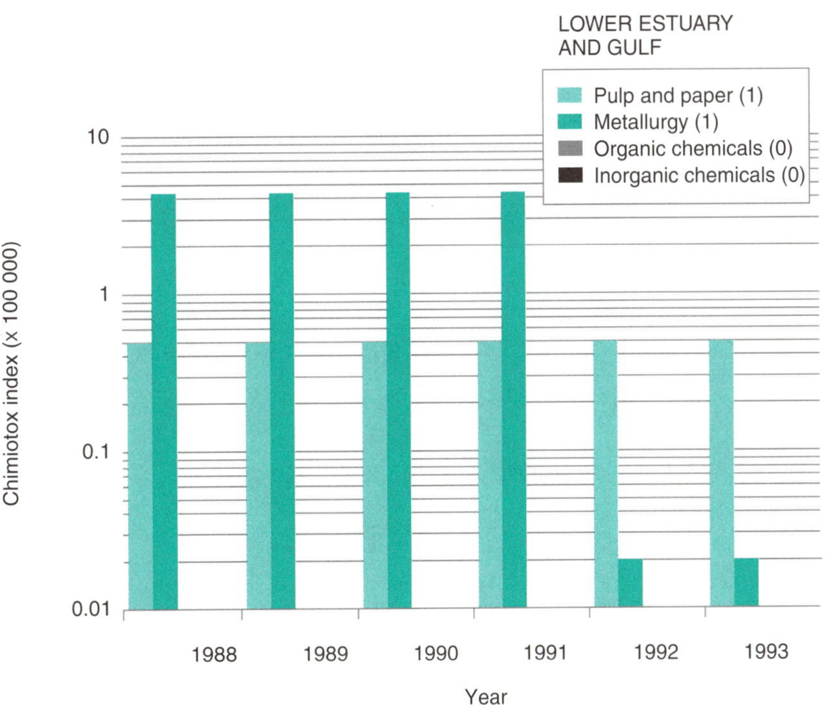

Note: Figures in brackets indicate the number of plants per industrial sector.

Source: Villeneuve, 1994.

Potential Ecotoxic Effects Probe (PEEP)

Definition

The PEEP index evaluates and compares the toxic potential of industrial effluent by means of a series of bioassays on organisms belonging to three trophic levels (bacteria, algae and micro-crustaceans). It incorporates several aspects of toxicity and is used to compare the potential toxic effects of industrial effluent on three types of very different organisms. The PEEP is also used to compare the toxic effects of different types of industries (pulp and paper, metallurgy, organic chemicals and inorganic chemicals), different plants, and different effluent from the same plant.

The PEEP index takes the form of a single value on a logarithmic scale, thereby allowing for rapid, unambiguous identification of those plants with the greatest toxic potential. Moving from one PEEP unit to the next represents a 10-fold increase in toxic potential. A value of 7 or more indicates a very high toxic potential. In magnitude, the scale is similar to the Richter scale used to evaluate the strength of seismic activity.

PEEP indexes for 49 of the 50 targeted plants studied were calculated using measurements taken between 1989 and 1992, and broken down by industrial sector: pulp and paper (15 mills), inorganic chemicals (11 plants), metallurgy (12 plants), and organic chemicals (11 plants). Results are shown for the four hydrographic regions of the River.

Limitations

The PEEP index provides no indication as to the potential toxicity of the aquatic environment. The effects measured by the PEEP incorporate overall ecotoxic phenomena, but provide no information on the nature of the elements responsible for the effects observed.

PROBLEM OR RISK–PRONE SECTORS

All river sections, to varying degrees: both the Fluvial Section and Fluvial Estuary in their entirety; the upper portions of the Upper Estuary and Saguenay River, in particular; and specific areas of Lower Estuary and Gulf.

Results

According to the results obtained by industrial sector, the highest toxicity values are presented by pulp and paper and inorganic chemicals, with values of 7.8 and 7.6 PEEP units (cumulative toxic loads), respectively. The relative contribution of each industrial sector to the sum of the toxic loads of the 49 effluents considered is 57.1% for pulp and paper, 39.1% for inorganic chemicals, 3.7% for metallurgy (6.6 PEEP units) and a mere 0.1% (5.0 PEEP units) for organic chemicals.

According to the results obtained by river sector, the relative contribution of each sector of the River to the total toxic load from the 49 effluents breaks down as follows: 42.9% (7.7 PEEP units) for the Fluvial Section, 30.4% (7.5 PEEP units) for the Fluvial Estuary, 10.7% (7.1 PEEP units) for the Upper Estuary and Saguenay, and 16.0% (7.2 PEEP units) for the Lower Estuary (see Figure 3.11.2). The stretch between Montreal and Varennes and the Fluvial Estuary between Trois-Rivières and Île d'Orléans are characterized by the combined presence of toxic effluent from all four industrial sectors identified.

Along the Fluvial Section there are 27 plants belonging to the four industrial sectors studied, distributed as follows: pulp and paper (2); inorganic chemicals (9); metallurgy (6); and organic chemicals (10). In the Fluvial Section, the Lake Saint-Pierre region receives the highest toxic load, 7.6 PEEP units. The effluent from the five plants in that region accounts for 83.2% of the toxic load from the main course of the River and 35.7% of the total toxic load from the 49 plants studied. The greatest input was from the Tioxide Canada Inc. plant, which alone accounted for 32.1% of the total toxic load from the 49 plants studied.

The six plants located in the first two regions of the Fluvial Section (Beauharnois Canal and Lake Saint-Louis) released toxic loads equivalent to PEEP values of 5.7 and 4.9, respectively. This input represents 0.4% and 0.1%, respectively, of the total toxic load from the 49 plants studied. For the Montreal–Varennes region, the 16 plants characterized discharge a toxic load of 6.9 PEEP units, or 6.7% of the total toxic load from the 49 plants studied. This high input comes primarily from the Kronos Canada Inc. plant, which alone accounts for 6.6% of the overall toxic load.

Along the Fluvial Estuary there are ten plants from the four industrial sectors studied: six for pulp and paper, one for inorganic chemicals, two for metallurgy, and one for organic chemicals. The region receives the second greatest toxic load, at 7.5 PEEP units or 30.4% of the overall toxic load from the 49 plants studied. The highest input comes from the effluent of the Canadian Pacific Forest Products Ltd. pulp and paper mill, which contributes 26.4% of the total toxic load from the 49 plants studied.

FIGURE 3.11.2

PEEP of cumulative loads from 49 of the 50 industrial plants targeted by the St. Lawrence Action Plan, by hydrographic region (1989-1992)

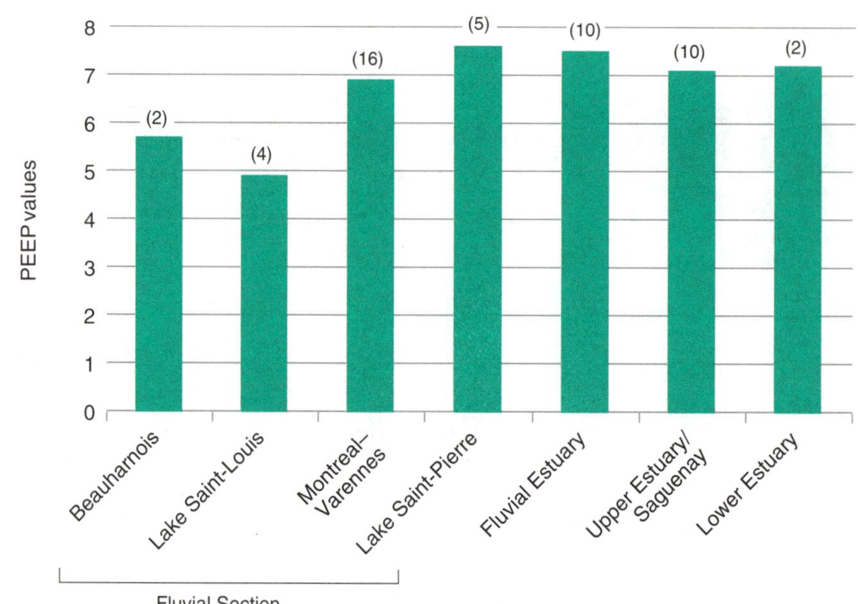

Note: Figures above bars indicate the number of plants.

Source: Based on data from Costan and Bermingham, 1993.

In the Upper Estuary/Saguenay sector there are ten plants, of which eight are located on the Saguenay River. The plants are divided among three industrial sectors: pulp and paper (6 mills), metallurgy (3 plants) and inorganic chemicals (1 plant). This river sector receives a toxic load of 7.1 PEEP units, or 10.7% of the overall toxic load from the 49 plants studied.

Inputs from the Stone-Consolidated Inc. Port-Alfred Division pulp and paper mill in La Baie are the region's highest, contributing 8.5% of the overall toxic load from the 49 plants studied.

Along the Lower Estuary there are two plants at Baie-Comeau, one pulp and paper mill, and one metallurgy plant. The greatest input comes from the QUNO Corporation pulp and paper mill, which alone accounts for 16.0% of the overall toxic load from the 49 plants studied.

Information Supplement

TOXIC SUBSTANCES RELEASED BY INDUSTRY: OILS, GREASES AND HEAVY METALS ARE STILL IN THE LEAD

Industries release a whole range of toxic substances into the River, and the final effluent from a single plant may contain several different types. Two effluents may also contain the same substances, but in differing concentrations and proportions. Chimiotox unit results from 1988 to 1993 show that oils and greases, and heavy metals (mercury, arsenic, chromium, nickel, silver and beryllium) are way ahead, with 32% and 29%, of the plants' entire toxic inputs. Other metals, including aluminum, iron and manganese, follow (9.8%), then PCBs (polychlorinated biphenyls) with 1.4%, PAHs (polycyclic aromatic hydrocarbons), 0.4%, and phthalates, 0.3%.

Between 1988 and 1993, source-reduction systems for contaminants or pre-discharge treatment processes for industrial wastewater were implemented by a number of industries. Aluminum smelters, for instance, now use clean processes (prebaked anodes) instead of polluting processes (Söderberg vats), and this has, for all intents and purposes, led to the elimination of toxics like PAHs.

The relative input of various substances to the overall toxicity of industrial effluent from the 50 plants did not radically change from 1988 to 1993. In 1993, heavy metals and oils and greases still headed the list of toxic substances discharged, but PAHs, PCBs and phthalates had almost completely disappeared (phthalates are generated by the petrochemical industry, where they are used to shape and mould plastics). Other highly toxic substances, such as dioxins and furans, accounted for 3% of the toxicity. Substantial (89%) reductions were posted for non-chlorinated phenols. Some toxic substances (such as elemental phosphorus, sulphides and chlorine) did not decrease, but their contribution to the toxicity of the effluent remains low, at 5%.

Reductions in discharges between 1988 and 1993 are shown in Figure 3.11.3 and illustrated by industrial sector. The 49 plants whose effluents underwent characterization as part of the St. Lawrence Action Plan reduced their liquid toxic waste by 74% over the period 1988-1993.

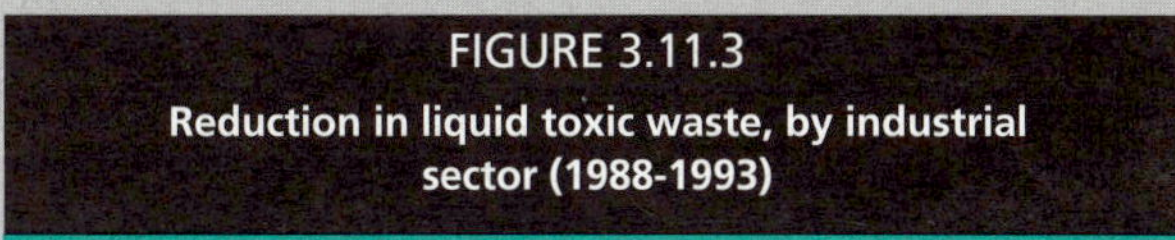
FIGURE 3.11.3
Reduction in liquid toxic waste, by industrial sector (1988-1993)

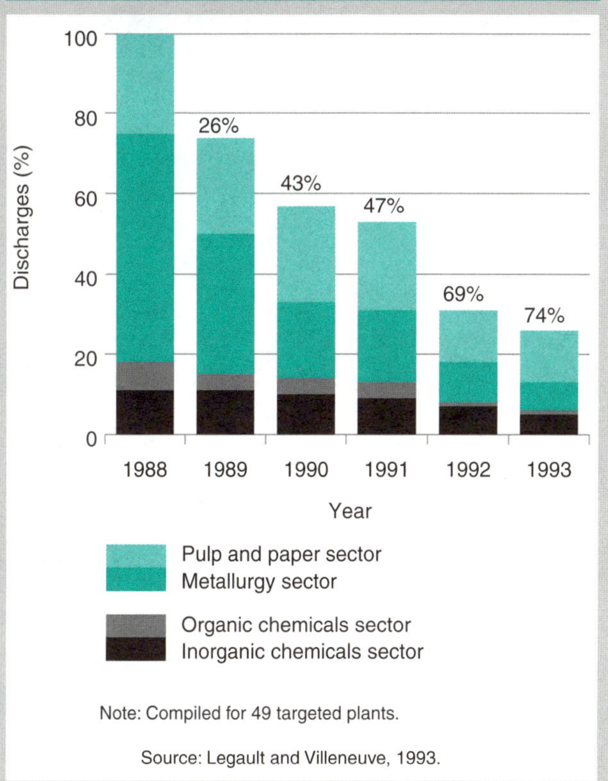

Note: Compiled for 49 targeted plants.

Source: Legault and Villeneuve, 1993.

Source: Bouchard, 1993.

INDUSTRIAL WASTEWATER DISCHARGES

Main Findings

- The measurements currently available relate solely to the 50 Action Plan targeted plants. Of these plants, 24 discharged their effluent directly into the River and are located between Cornwall and Quebec City; in 1991, they were responsible for approximately 9% of the toxic load in the River. The 50 targeted plants belong to the pulp and paper, metallurgy, organic chemicals and inorganic chemicals sectors. The two upstream sections of the River are subject to the greatest inputs, followed by the upper parts of the Upper Estuary and Saguenay, and the Baie-Comeau area of the Lower Estuary. There are 6300 smaller facilities along the River, only some of which are connected to a municipal sewer system.
- Liquid toxic waste from the 49 priority plants fell from 5.2 to 1.3 million Chimiotox units between 1988 and 1993.
- In 1992, the volume of process water released by these industries was over 1.4 million cubic metres per day (56% from the pulp and paper mills, and 35% from metallurgical plants). This discharge contained 272 tonnes of suspended solids. In 1993, industrial effluents still contained numerous toxic substances (the Chimiotox index considers 124 substances): oils and greases, heavy metals and other metals (including aluminum, iron and manganese), PCBs, resin acids and fatty acids, etc.
- Measurements made between 1989 and 1992 with the Potential Ecotoxic Effects Probe (PEEP) on organisms exposed to industrial effluent and not to the natural environment show higher potential toxicity indexes for the pulp and paper and inorganic chemicals sectors. Of the four river sectors, the Fluvial Section contributes the most to the total toxic load of the 49 plant effluents, with 43% or 7.7 PEEP units.

Relative Importance

▲ Industrial wastewater discharges tops the list of the eight most influential characteristics for the St. Lawrence as a whole.

Recommendations on Monitoring

- The Chimiotox and PEEP indexes would describe the state of this characteristic more accurately if they were applied to a larger number of industrial plants and measured other important parameters based on knowledge of the environment.
- Priority effluent should be characterized periodically with the indicators available. Moreover, while the PEEP incorporates the overall antagonistic, additive or synergistic phenomena of the toxic substances in the effluent, another synergy may occur when the effluent meets the water of the receiving body. It is therefore strongly recommended that a similar scale be developed for the receiving body. By weighting this index according to the location of the outfall and the specific features of the receiving body, these toxic discharges currently measured in the final effluent could be monitored in the environment.

3.12

COMMERCIAL FISHING

Fisheries and Oceans

Background

Fifteen years ago, the commercial fishery began to decline. For various reasons which are still not understood clearly, some species of fish have virtually disappeared, whereas the stocks of other species appear to have been permanently destabilized (see information supplements entitled, "Fish species at risk in the St. Lawrence" and "The American eel in troubled waters").

Specialists are rarely able to pinpoint the causes of destabilization of a given stock of fish. Some of the suspected causes are a reduction in vital functions, such as reproduction or respiration, resulting from chronic exposure to toxic substances, physical destruction of spawning grounds or disturbances caused by sedimentation, artificial obstacles such as jetties or dams impeding the migration of adult fish, and overfishing. Whether a single cause or a combination of causes is involved, destabilization of stocks can permanently mark a species: with its numbers reduced, its age structure skewed and its habitats degraded, the population may be pushed toward extinction.

In most cases, gross landed volume is the only measurement tool available, so it has been selected here. Although this indicator is directly linked to the condition of biological resources, it has certain limitations. Furthermore, catch volumes are influenced by commercial factors such as changes in market prices, catch quotas and applicable legislation.

Catch data on commercial fishing for various freshwater and saltwater species are presented below; they cover different periods depending on the species considered.

Interpretation

The populations of some fish species, such as Lake sturgeon and Atlantic cod, continue to decline. The volume of freshwater landings is estimated to have decreased by half between 1945-1960 and 1985-1991. Other data, including information on commercial species, compiled from experimental fishing carried out over the past 20 years in the Quebec City sector, indicate that annual freshwater catches have dropped sharply there, from roughly 50 000 fish per year in 1971 and 1972 to 10 000 per year in 1991 and 1992. In the saltwater sector, data on four commercially-important groundfish species of the marine environment show an upward trend in total landed volume between 1979 and 1987. In 1992, however, the total catch represented 65% of the average volume recorded during this period, and cod landings were roughly equivalent to the lowest levels of the 1970s.

Fresh Water

Landings of certain species

Definition

The fish species listed in Figure 3.12.1 were the four most heavily harvested in fresh water during the 1986 to 1991 period: Brown bullhead, Yellow perch, Lake sturgeon and American eel. Two of these species, Lake sturgeon and American eel, are at risk.

Limitations

The landed value and type of fish caught are directly linked to economic and social factors.

Problem Or Risk-Prone Sectors

Varies with the species considered.

Results

Landings of Lake sturgeon and American eel fluctuate widely from year to year, sometimes by as much as 30%. Whereas peaks in catches of Lake sturgeon were observed in 1987 and 1990, American eel landings peaked in 1986 and 1990. Between 1986 and 1991, landings of Yellow perch fluctuated slightly. Brown bullhead catches declined steadily between 1986 and 1991, with a particularly steep drop in 1989. The landed volume for this species was twice as high in 1986 as in 1991.

Overall, the total volume of landings in the freshwater sector is estimated to have dropped by half between the 1940s and the 1980s, with the number of fishermen decreasing by 80% during this same period.

FIGURE 3.12.1

Annual landings of four freshwater fish species of the St. Lawrence (1986-1991)

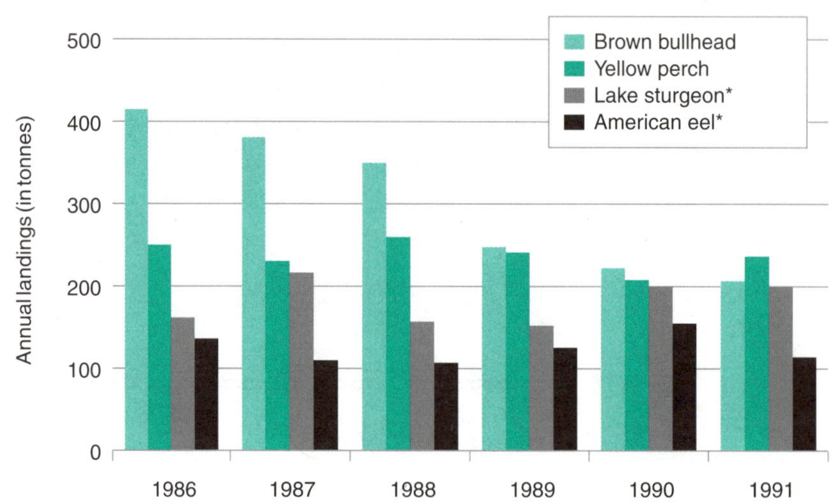

Note: *Species at risk in the St. Lawrence.

Sectors studied: Lake Saint-François, Lake Saint-Louis, La Prairie basin, Lake Saint-Pierre, Trois-Rivières and Quebec City sectors.

Source: Based on data from Johnson, 1991.

Salt Water

Landings of certain species

Definition

Listed in Figure 3.12.2 are the four pelagic and estuary species, according to Fisheries and Oceans classification, most heavily fished in the estuary during the 1984 to 1992 period: herring, Mackerel, Capelin and eel. The herring and eel are considered at risk.

Landings of four groundfish species most heavily fished from 1971 to 1992 are shown in Figure 3.12.3: Greenland halibut, Canadian plaice, redfish and Atlantic cod. Crustacean catches between 1970 and 1992 are shown for Pink shrimp (see Figure 3.12.4) and Northern lobster (see Figure 3.12.5).

Limitations

The landed value and type of species fished are directly linked to economic and social factors.

Problem or Risk-Prone Sectors

Upper Estuary and Saguenay | Lower Estuary and Gulf

Varies with the species considered.

Results

The herring has the highest landed volume; fluctuations in catches ranged from 2 to 48% between 1986 and 1991 (see Figure 3.12.2). Catches of the three other saltwater species fluctuate considerably from year to year, with Capelin fluctuating the most. Indeed, landings of this species increased dramatically (82%) in 1987, and then plummeted 79% in 1990. The largest catches of Mackerel and Capelin, respectively, occurred in 1988 and 1989. Landings of eel in the estuary average three times those in fresh water. Nonetheless, the American eel population appears to be in difficulty, and the decline in recruitment of this species, which plays a very important role in the aquatic communities of the Lake Ontario–St. Lawrence River ecosystem, points to problems in the years ahead (see information supplement entitled, "The American eel in troubled waters").

FIGURE 3.12.2

Annual landings of four pelagic and estuary fish species of the St. Lawrence (1984-1992)

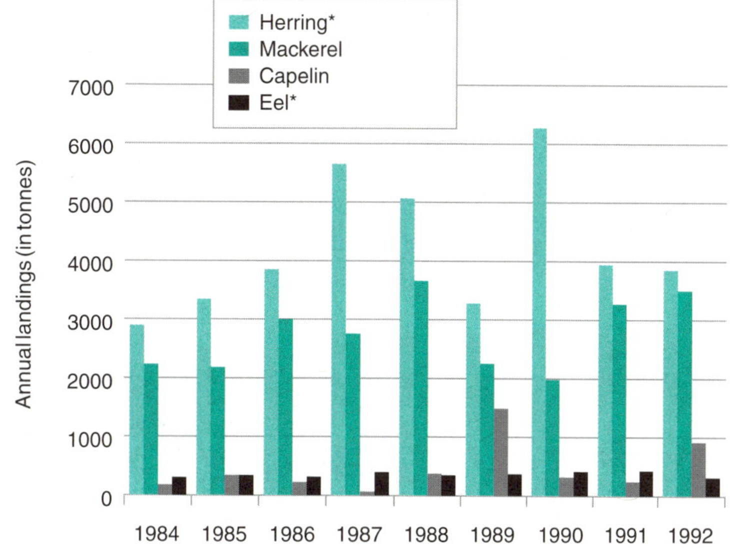

Note: * Species at risk in the St. Lawrence.

Sources: Based on data from Fisheries and Oceans, 1988, 1990, 1992a, 1993b.

Among groundfish species in the marine environment, Atlantic cod has always had the highest landed volume and value in Quebec. Catches of cod have always been highly variable, owing to the combined effects of numerous climatic, biological, socio-economic and technical factors. Trends in landings over the past 20 years show a sharp drop between 1971 and 1976, an upsurge between 1977 and 1981 after the 200-mile limit was established in 1977, relative stability over the following four years and a sharp decrease since 1985 (drop of 62% in seven years) (see Figure 3.12.3).

FIGURE 3.12.3
Annual landings of four groundfish species of the marine environment of the St. Lawrence (1971-1992)

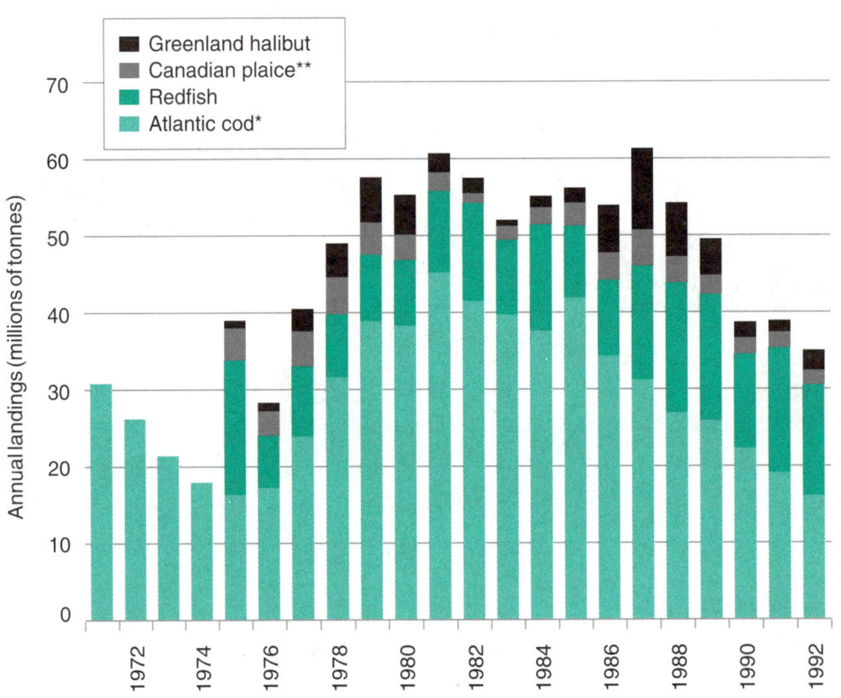

*Species at risk in the St. Lawrence.
**Includes Canadian plaice, Winter flounder, Witch flounder and Yellowtail flounder.

Note: Sector studied: downstream of Montmagny, including Îles de la Madeleine. Where no bar appears, the data were unavailable.

Source: Based on data from Fisheries and Oceans, 1993a.

In 1992, the cod catch fell to near the record lows of 1974 to 1976, and in 1993 the total allowable catch (TAC) for the Gulf was cut back substantially to 13 000 t from 43 000 t the year before. In the northern Gulf, the TAC was reduced from 35 000 to 3000 tonnes (Therriault 1993). These decreases reflect the growing scarcity of cod stocks in the Gulf.

Landings of redfish declined markedly in the 1970s, then increased slightly overall between 1981 and 1984. Toward the end of the 1980s, catches returned to the levels seen in 1975. Since 1980, landings have fluctuated widely from year to year, with an 11% drop in 1991 compared to 1992.

Landings of Canadian plaice during the period 1971-1992 exhibit three general trends: first, a steep drop between 1978 and 1982, followed by an upturn, reaching a peak in 1987. Since 1988, catches have declined steadily, dropping 60% in five years.

Greenland halibut landings appear to fluctuate over a four-year cycle. Two uptrends can be seen, the first between 1975 and 1979 and the second between 1983 and 1987. In 1987, catches reached the highest level in 18 years. Downturns in landings were recorded for the periods 1979-1983 and 1987-1991.

In 1992, landings of Greenland halibut began climbing again after four straight years of decline.

Overall, the declining volume of landings of three of these four species (see Figure 3.12.3) has accelerated since 1987. In 1992, cod landings made up 55% of the average landed volume posted between 1971 and 1992; Canadian plaice 61%; and Greenland halibut 72%. In 1992, the landed volume of redfish represented 120% of the average volume recorded for the species between 1975 and 1992.

After the shrimp fishery opened in 1965, the volume of landings remained fairly low until 1972, after which it rose steadily (see Figure 3.12.4), peaking in the early 1990s. Since 1991, the Sept-Îles and Anticosti Island sectors have posted declines, with the most dramatic drop occurring in 1992, when the total catch for the four sectors was down by 40% from 1991. Nonetheless, the 1992 landings were higher than the levels of the 1970s.

FIGURE 3.12.4

Annual landings of Pink shrimp, by management unit (1970-1992)

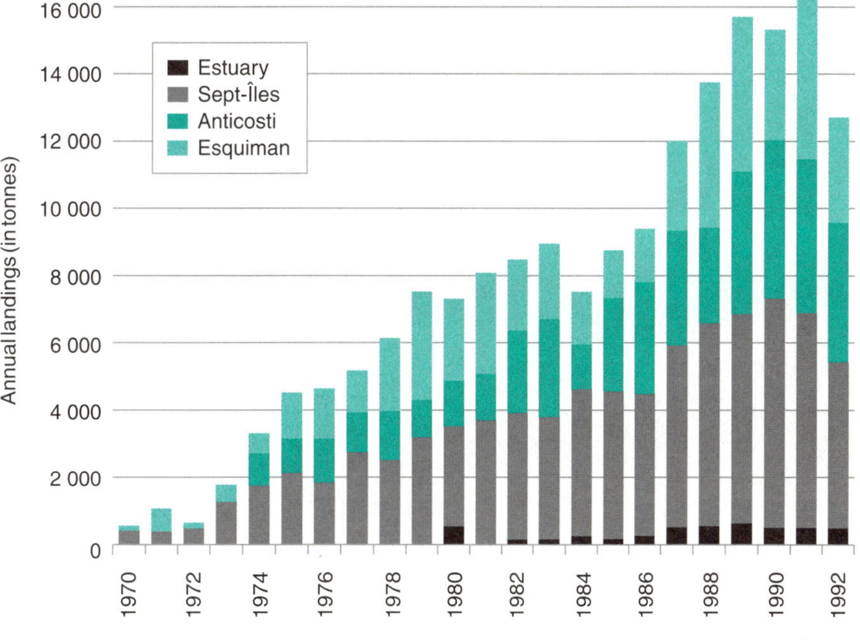

Note: Where no portion of bar appears, the data were unavailable.

Source: Based on data from Savard, 1993.

Lobster catches were almost two times lower (see Figure 3.12.5) in the 1970s than in the 1980s. Landings of this species have been on the rise since 1985, although the reason for this upward trend remains unknown. Between 1984 and 1992, landings for the two sectors considered rose by 100%.

FIGURE 3.12.5

Annual landings of Northern lobster in the Îles de la Madeleine and Gaspésie (1970-1992)

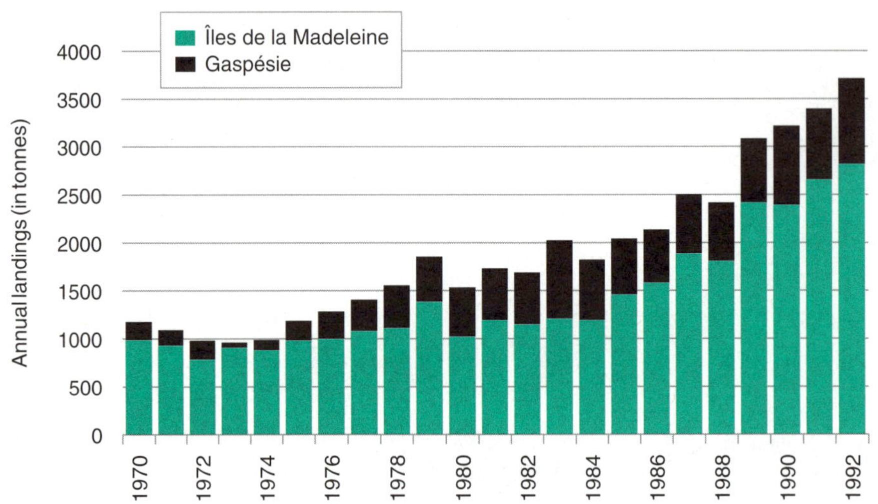

Source: Based on data from Fisheries and Oceans, 1988, 1990, 1992a, 1993b; BSQ, 1990.

Information Supplement

FISH SPECIES AT RISK IN THE ST. LAWRENCE

Ten species of fish are considered at risk in the St. Lawrence. Most are migratory species that frequent very different parts of the ecosystem during their life cycle (Rainbow smelt, Atlantic tomcod and Striped bass), or which leave the St. Lawrence completely and return after an absence of several years (American eel, Atlantic salmon and American shad). All these species have been or still are intensely harvested by commercial and sport fishermen. This list is not exhaustive; sedentary species that are not fished, such as the Copper redhorse, are also at risk, but for different reasons.

Lake sturgeon (*Acipenser fulvescens*). Recruitment in this species has declined due to the low abundance of spawners and limited spawning grounds. Commercial overfishing and a high natural mortality rate associated with the poor quality of habitats have contributed to its decline.

Atlantic sturgeon (*Acipeser oxyrhynchus*). Commercial landings currently consist only of fish that have not reached sexual maturity. Wide variations in catch levels suggest that the St. Lawrence stock is destablilized, due mainly to reasons associated with dredging for the St. Lawrence Seaway. Other factors include local overfishing, the impacts of development work for Expo 67, and more efficient fishing gear.

American eel (*Anguilla rostrata*). The causes of the decline in this species are not known, although there are several possibilities. Pollution in certain stretches of the River crossed by the eels while migrating toward the Atlantic Ocean may have caused the deaths of some spawners. Judging from catches at the Quebec Aquarium, the decrease in abundance of migrating adults has been gradual. By contrast, the number of juveniles heading seaward from the eel ladder at the Moses Saunders dam in Cornwall has declined dramatically.

American shad (*Alosa sapidissima*). At present, this anadromous species is caught only occasionally in the herring fisheries of the Upper Estuary in the spring, and in early summer in the Montreal area. The many alterations in the river bed caused by the construction of power plants (Beauharnois, Rivière-des-Prairies and Île des Moulins) may have cut the species off from its traditional migration routes to spawning grounds, some of which are located in Lake Ontario.

Northern pike (*Esox lucius*). Although this predator is ubiquitous in the warmer waters upstream from Lake Saint-Pierre, it is considered at risk downstream from there. The species' abundance is relatively low, and it is feared that encroachments and shoreline work (filling) have drastically reduced its preferred habitat.

Atlantic tomcod (*Microgadus tomcod*). This species is common in the intertidal communities of the St. Lawrence. Its spawning area has shrunk and is now limited to two rivers, the Sainte-Anne and the Batiscan. Tomcod typically spawn in winter; the larvae migrate downstream through the saline transition zone and the intertidal areas in the lower reaches of the River. Several factors are believed to underlie its recent decline: the disappearance of two successive cohorts (1984-1985 and 1985-1986) may have destabilized the population, while alterations to the flow pattern of the River (construction of wharves that impede migration) and the loss of spawning grounds owing to the contamination of certain tributaries may have added to the tomcod's woes.

Rainbow smelt (*Osmerus mordax*). Of the four distinct Rainbow smelt populations that inhabit the ecosystem, only one is at risk – the population living along the south bank of the Upper Estuary. The collapse of the Rainbow smelt population in the Upper Estuary may be linked to viral diseases, overfishing during the spawning period and habitat degradation in the few tributaries used by the species for spawning (the Boyer River, in particular), as well as the overall deterioration in water quality.

Striped bass (*Morone saxatilis*). A characteristic resident of the turbid waters in the upper part of the Upper Estuary, the Striped bass formerly moved upstream to overwinter in the Lake Saint-Pierre region. However, the construction of the Seaway and four other major works, including the islands created for Expo 67, may have played an indirect role in the species' near-extinction. Up until 1966, there was a thriving commercial fishery for Striped bass.

Anadromous Brook charr (*Slavelinus fontinalis*). Owing to the sketchy knowledge of this species' habits, its decline cannot be explained easily. Brook charr appear to remain near the estuary of their river of origin, to which they return to spawn upon reaching maturity. Conditions around the mouths of different rivers may explain why Brook charr numbers have declined; problems include physical degradation of habitat, poaching, commercial fishing and overharvesting by sport fishermen.

Atlantic salmon (*Salmo salar*). A widespread decrease caused by overfishing of Atlantic salmon stocks in pelagic environments (Labrador Sea) and harvesting of spawners by the inshore commercial fishery have brought about an overall reduction in the abundance of Atlantic salmon. The population is nonetheless starting to make a comeback, and thanks to judicious stocking and strict monitoring of fishing activity, the species may well be able to rebuild.

Source: Based on data from Fisheries and Oceans, 1991.

Information Supplement

THE AMERICAN EEL IN TROUBLED WATERS

After migrating to the St. Lawrence as elvers, juvenile eels begin to move upstream. While some of them settle in the main river bed, the majority swim up tributaries, where they remain until reaching sexual maturity between the ages of 10 and 12. A large proportion of juveniles that undertake the upstream migration, which lasts at least four years, end up in Lake Ontario. A counting system at the Moses Saunders dam near Cornwall has recorded the alarming downturn in the number of eels migrating upstream: whereas 1 293 470 juveniles were counted in 1983, there were only 11 533 in 1992. A drop in recruitment of this magnitude points to a decline in the commercial fishery of the St. Lawrence beginning in 1996, which is when eels from the increasingly weaker cohorts that migrated upriver between 1986 and 1992 will begin their voyage back to the Atlantic Ocean.

To what can the current decline be attributed? Looking back over time, it is conceivable that the low recruitment observed at the Moses Saunders dam between 1986 and 1992 resulted from reduced spawning by the adult fish that migrated downstream between 1980 and 1986. The average annual commercial catch of St. Lawrence eels (River and Lake Ontario) fell from 715 tonnes for the period 1975-1981 to 537 t for 1984-1991, a drop of 25%. No data is available, however, on fishing effort for this species over its entire geographic range and spanning several years. As such, we can not conclude that overfishing in the St. Lawrence is the principal cause of the drastic decline begun in 1986.

Two other hypotheses focus on environmental contamination by toxic substances and physical modifications to eel habitat. Cases of mass mortality in the past were blamed on damage to eels' gills caused by mechanical factors (turbines of hydro-electric dams, fishing gear, vessels and predators) or by heavy pollution in the bodies of water where the eels spent much of their life cycle. However, the eels with lesions on their gills were found to be no more contaminated than the healthy eels examined; hence, their death can not be attributed to chronic exposure to toxic substances.

No real link has been identified between the start of the population decline and contamination of eels by chemicals or habitat disturbances resulting mainly from construction of the Seaway. A relationship of this type seems unlikely, given the lengthy time lags between these events and the observed decrease in eel numbers. It is nonetheless possible that chemical and habitat disturbances played a role in the decline, but that their impact was delayed by a factor that remains unknown.

A similar population decline has been noted in the two eel species that spawn in the Sargasso Sea, the American eel and the European eel. This would seem to indicate that a more global cause may be involved, such as oceanic changes that affected both species. Oceanic changes would influence migration and spawning by adults, as well as the transport, survival, growth and development of eel larvae during their migration toward the continent. After eel larvae are born in the Sargasso Sea, they drift landward on the Gulf Stream. In the 1980s, a slowing of the Gulf Stream was observed. A slower current could hinder the transport of larvae and cause the declines noted in both eel species. Furthermore, the physical processes linked to larval transport may have been altered by the global warming trend and associated oceanic changes in the Northwest Atlantic.

Whatever the case may be, a clear link has not been established between any of these possible causes and the decline in eel populations. These causes are difficult to verify, given the eels' complex life cycle. The decrease in eel numbers is likely due to a combination of factors acting in concert. While oceanic conditions may be the primary cause, the other three causes (contamination by toxic substances, habitat alteration and excessive commercial fishing) may have made matters worse by reducing the number of eels that return to the Sargasso Sea to spawn. Little information is available for evaluating the synergy of the different causes and their respective contribution to the decline in eel numbers.

Researchers believe that the decrease in eel numbers may affect the entire Northwest Atlantic stock. The species' geographic range encompasses the St. Lawrence, the Great Lakes and most of the tributaries along the east coast of North America. The St. Lawrence is home to a significant part of the stock's female contingent, which are the only eels that migrate upriver. If, as feared, they disappear from the contingent that reaches the Sargasso Sea over the next decade, the entire Northwest Atlantic stock, which is considered a single population, may be destabilized. Although catches of American eels in the St. Lawrence have fluctuated widely before, some experts see new, worrisome signs in the current situation.

Sources: Castonguay et al., 1994a; 1994b.

COMMERCIAL FISHING

Main Findings

- Between 1984 and 1992, landings of lobster in the Îles de la Madeleine and the Gaspé rose by 100%.
- The total volume of landings of Brown bullhead, Yellow perch, Lake sturgeon and American eel in the freshwater sector is estimated to have dropped by half between the 1940s and the 1980s. This decline has worsened since 1986.
- In the marine sector, commercial fishing for the most harvested groundfish species (Greenland halibut, Canadian plaice, redfish and Atlantic cod) has declined rapidly since 1987. Landings of cod have decreased sharply since 1985, dropping by 62% in seven years. Between 1984 and 1992, landings of pelagic and estuary fish species like herring, Mackerel, Capelin and eel fluctuated considerably from year to year, but declined the most beginning in 1989. Landings of eel in the estuary are an average three times higher than those in fresh water.
- The volume of shrimp landings rose steadily beginning in 1972, peaked in the early 1990s, and declined the most dramatically in 1992. In the 1980s, lobster catches were almost double what they had been in the 1970s. Landings of lobster have been on the increase since 1985, rising by 100% between 1984 and 1992.
- Some commercial species, like American shad and Atlantic sturgeon, underwent declines in the past from which they never recovered. Rainbow smelt and Atlantic cod stocks are affected, and there are signs of a gradual decline in Snow crab catches, giving even greater cause for concern. Furthermore, commercial landings of Atlantic sturgeon now consist solely of individuals that have not reached sexual maturity.

Relative Importance

- Commercial fishing, as a hybrid characteristic, ranks sixth out of the seven most influenced characteristics, and the seventh most influencial of the eight characteristics for the St. Lawrence as a whole.

Recommendation on Monitoring

- The volume of landings per unit of effort would be more useful for assessing the status of this characteristic, since it includes a strong biological component: the condition of resources. Nonetheless, the yield per unit of effort is also considered a measure of profitability in the industry, and is influenced by management policies and decisions.

3.13

SPORT HUNTING AND FISHING

Canadian Wildlife Service, L.-G. de Repentigny

Background

The continuing appeal of sport hunting and fishing make this a useful characteristic to consider in monitoring the condition of the River, especially in view of its direct relationship with the status of biological resources (see information supplement entitled, "Lead shot contamination of waterfowl along the St. Lawrence"), natural environments and protected species, and with access to the banks and the River itself. Furthermore, sport fishing is linked to the commercial fishery because of competition for some of the same species, like Atlantic salmon (see information supplement entitled, "The return of the Atlantic salmon").

We do not have all the information needed to assess the state of sport hunting and fishing or the patterns of the past 15 years. For now, three indicators have been selected as useful for describing several aspects of this characteristic: waterfowl harvests, catches of certain species of sport fish, and restrictions on consumption of fish.

Exposure to persistent organic and inorganic toxic substances from eating contaminated fish poses a risk to human health, and restrictions on consumption can affect sport fishing.

Interpretation

The number of waterfowl taken is recorded every year for Quebec as a whole. Between 1988 and 1991, the waterfowl harvest in Quebec increased by 12%, from 511 200 to 574 000. In 1991, the harvest shrank by 20% after three years of constant growth. The average annual waterfowl harvest in Quebec between 1988 and 1991 was an estimated 519 500 birds: 423 300 ducks and 96 200 geese and Canada geese. Since almost all the Greater snow geese and 65% of the ducks harvested in Quebec are taken along the St. Lawrence, the average annual harvest of waterfowl connected with the St. Lawrence can be estimated at 371 300. Since 1988, however, there has been a steady increase in the proportion of geese and Canada geese in the total annual harvest of waterfowl. In

1988, geese and Canada geese made up 13% of the provincial harvest, compared with 22% in 1991, a rise of 9% in four years.

Between 1977 and 1981, the Fluvial Section and the Upper Estuary accounted for the largest share of the average annual harvest of waterfowl. For the River corridor as a whole, the most widely harvested group of species was dabbling ducks, followed by diving ducks, geese and Canada geese, and then sea ducks. The Black duck, the Mallard, the Green-winged teal and the Greater snow goose were the most heavily harvested species in the four sectors combined, accounting for nearly 50% of the total harvest.

Just as with the commercial fishery, some popular species of fish targeted by the sport fishery have all but disappeared, and others are currently at risk, notably the Northern pike, Atlantic tomcod and Rainbow smelt. Atlantic salmon, which was overfished on the high seas in the past, is now recovering. The number of commercial salmon fishermen has fallen steadily over the past few years.

In the mid-1980s, the Cornwall–Sorel sector had the highest total volume of landings. For the River corridor as a whole, pike, Yellow perch, Walleye and Atlantic tomcod were the most heavily fished species. Approaching the estuary, the number of migratory species targeted by sport fishermen rises.

Restrictions on consumption have generally been eased since 1985. Between 1992 and 1993, the guidelines did not change, except with respect to Walleye and Smallmouth bass in the Trois-Rivières–Quebec City sector, where the maximum number of meals per month was increased from two to four, and Yellow perch in Lake Saint-Louis, where the maximum was raised from two meals per month to eight. For this same period, Northern pike in the Lake Saint-Louis sector and Walleye in Repentigny–Sorel were subject to the most stringent restrictions.

Sport Hunting

Waterfowl harvest: Number and species of birds taken

Definition

Waterfowl hunting is a traditional activity that is still carried out along the River. From 1977 to 1981, a survey was conducted of waterfowl harvesting along the River. These data were assembled based on the four hydrographic regions of the St. Lawrence. Four groups of birds are covered – geese and Canada geese, dabbling ducks, diving ducks and sea ducks.

Limitations

The data are from several years ago and do not take seasonal fluctuations into account.

Results

Problem Or Risk-Prone Sectors

FLUVIAL SECTION	FLUVIAL ESTUARY
UPPER ESTUARY AND SAGUENAY	LOWER ESTUARY AND GULF

Varies with the species considered.

Between 1977 and 1981, the average annual harvest of waterfowl along the St. Lawrence River totalled 372 300 birds (see Figure 3.13.1). Roughly 47% of the harvest takes place along the Fluvial Section, the most urbanized part of the St. Lawrence, followed by 24% in the Upper Estuary, 17% in the Fluvial Estuary and 12% in the Lower Estuary and the Gulf.

During this period, the waterfowl harvest was made up of 31 species: 9 dabbling ducks (chiefly the Black duck, Mallard, Green-winged teal and Blue-winged teal), 11 diving ducks (chiefly the Lesser scaup, Common goldeneye, Greater scaup and Ring-necked duck), 8 sea ducks (chiefly the Surf scoter, Black scoter, Oldsquaw, Common eider and White-winged scoter) and 3 anserine species (chiefly the Greater snow goose and Canada goose). The Black duck, Mallard, Green-winged teal and the Greater snow goose were the four most heavily harvested species in the four sectors combined, accounting for nearly 50% of the total harvest.

FIGURE 3.13.1
Average annual waterfowl harvest along the St. Lawrence, by river sector (1977-1981)

Lower Estuary/Gulf (45 166)
Upper Estuary (87 841)
Fluvial Section (174 427)
Fluvial Estuary (64 849)

Source: Based on data from Lehoux et al., 1985.

Dabbling ducks were the most heavily harvested group in all sectors of the River, with 58% of the harvest in the Fluvial Section, 57% in the Fluvial Estuary, 47% in the Upper Estuary and 44% in the Lower Estuary and Gulf. Along the Fluvial Section and the Fluvial Estuary, diving ducks ranked second, making up 28% and 29% of the total harvest in each these sectors. In the Upper Estuary, the group of geese and Canada geese came in second place, with 38% of the birds taken in this sector, whereas in the Lower Estuary and Gulf, sea ducks ranked second, with 26% of the total.

More than half of the dabbling ducks and diving ducks are harvested in the main course of the River. Sea ducks are hunted primarily in the Fluvial Section (39%) and the Lower Estuary and Gulf (33%). The harvest of geese and Canada geese is centred in the Upper Estuary (62%).

Sport Fishing

Catches of certain species

Definition

The landings of game-fish species for which data are available cover six species fished in the Fluvial Section and the Fluvial Estuary. These catches illustrate the extent of freshwater sport fishing in the 1980s.

Limitations

Data on catches by sport fishermen are incomplete and were compiled from studies conducted mainly between 1983 and 1985. They deal only with sport fishing activities in fresh water, between Cornwall and the eastern tip of Île d'Orléans, specifically landings of bass, pike, Walleye, Yellow perch, Atlantic tomcod and Rainbow smelt. Three of these species – Northern pike, Atlantic tomcod and Rainbow smelt – are considered to be at risk.

Problem Or Risk-Prone Sectors

Varies with the species considered.

*Species at risk: Cornwall to Grondines (Northern pike), Sorel to Île d'Orléans (Atlantic tomcod), Grondines to Île d'Orléans (Rainbow smelt).

Results

In the 1980s, sport fishermen caught approximately 3260 tonnes of fish per year in the freshwater part of the St. Lawrence (see Figure 3.13.2). In the mid-1980s, the main game species caught were, in order of importance, pike (including Muskellunge), Yellow perch, Walleye, and Atlantic tomcod. The catch volume decreases seaward. The bulk of the harvest takes place in the Montreal (or Cornwall–Sorel) sector, with 80% of the annual freshwater catch. Here, pike are caught most often, followed closely by Yellow perch and Walleye. The Sorel–Grondines sector accounts for 18% of the annual catch, with tomcod heading the list, followed by Walleye. As regards the Grondines–Île d'Orléans sector, where the sport fishery is not well developed, tomcod is the main species caught (2% of the harvest).

FIGURE 3.13.2
Estimated catches of the main game-fish species harvested in fresh water in the St. Lawrence, by river sector (1983-1985)

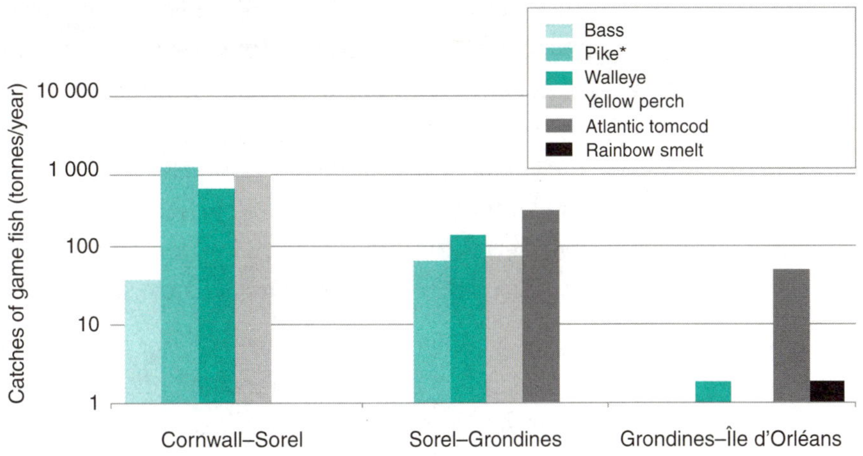

Note: Only the most heavily fished species are shown for each sector.

*Including Muskellunge.

Source: Based on data from Mailhot, 1989.

Approximately 94% of the pike taken in the study area comes from the Cornwall–Sorel sector and 6% from the Sorel–Grondines sector. In the case of Yellow perch, 92% of the catch originates from Cornwall–Sorel and 8% is from Sorel–Grondines. Some 80% of the Walleye is taken in the Cornwall–Sorel sector. Most of the tomcod, or 86%, is caught in the Sorel–Grondines region, with the balance coming from Grondines–Île d'Orléans. Rainbow smelt are caught in the Grondines–Île d'Orléans sector.

Sport Fishing

Restrictions on consumption of fish:

Maximum number of meals recommended per month for various species

Definition

Research conducted in recent years on freshwater fish species between Cornwall and the eastern tip of Île d'Orléans has identified some of the contaminants to which the species are exposed: mercury, polychlorinated biphenyls (PCBs), DDTs, hexachlorobenzene (HCB), dieldrin, dioxins and furans. Fish accumulate variable quantities of contaminants in their flesh, depending on where they are captured, their size and their position in the food chain. Large fish and predatory species are most likely to contain high concentrations of contaminants.

Consumption guidelines indicate the maximum number of meals recommended per month (values for large fish), and vary among species type and river sectors. These restrictions reflect potential contaminant levels in species and concern for human health. Changes in this indicator over time show the evolution in the overall health of the environment. Restrictions may either reduce or increase the level of sport fishing activity. Data are shown for five freshwater species by river sector for 1992 and 1993. Consumption guidelines are given only for large fish, since they are subject to the most stringent restrictions.

Each sector of the River corridor contains a number of sites. For each sector, the number of meals recommended is dictated by the sites that have the most severe restrictions. A meal consists of 230 grams (8 ounces) of fresh, uncooked fish.

Limitations

Each consumption restriction is based on the overall contamination of fish and must take into account all species eaten during the month. Data on restrictions are not always available for all species or all sectors of the River corridor.

Problem Or Risk-Prone Sectors

FLUVIAL SECTION | FLUVIAL ESTUARY

The most stringent restrictions apply to Repentigny–Sorel and Lake Saint-Louis in the Fluvial Section.

Results

Consumption guidelines did not really change between 1992 and 1993, apart from certain cases in which restrictions were eased. In the Trois-Rivières–Quebec City sector, consumption recommendations for Walleye and Smallmouth bass were raised from two to a maximum of four meals per month. Restrictions on Yellow perch from Lake Saint-Louis were also eased from two meals per month in 1992 to a maximum of eight meals per month in 1993.

In 1992 and 1993, the most severe restriction applied to Walleye in the Repentigny–Sorel sector (maximum of one meal per month) and to Northern pike in Lake Saint-Louis (maximum of one meal per month).

Based on a comparison of available data for the five sectors, the three farthest upstream are subject to the most stringent restrictions (in descending order, Lake Saint-Louis, Lake Saint-François and Repentigny). There are few data on consumption restrictions for the Trois-Rivières and Quebec City sectors.

FIGURE 3.13.3

Maximum number of meals recommended for the consumption of five species of game fish, by river sector (1992-1993)

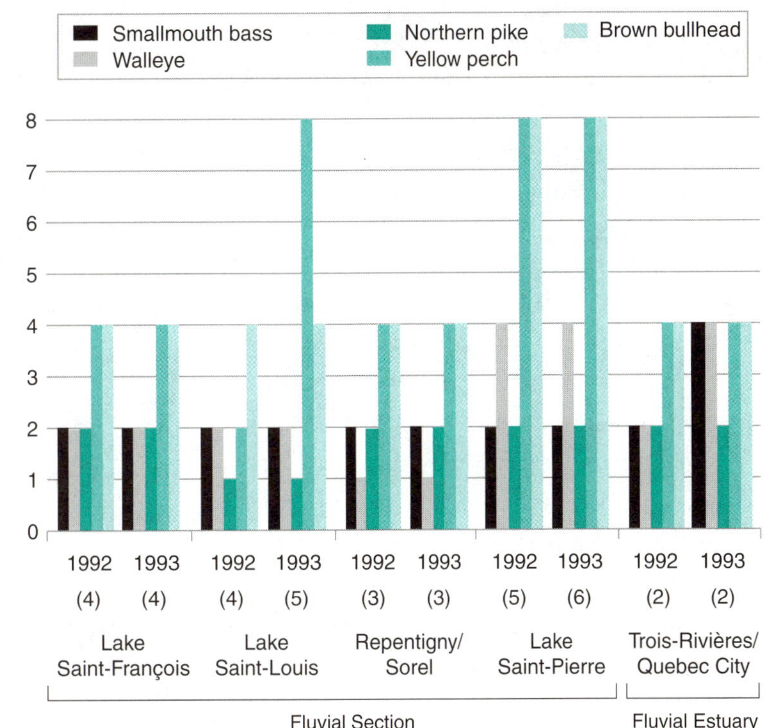

Note: A meal consists of 230 g of fresh, uncooked fish.

Figures in brackets indicate the number of fishing sites surveyed by sector. The maximum number of meals per month is based on the most severe restriction applying to each sector.

The number of specimens caught is not available in all cases.

Source: Based on data from MSSS and MENVIQ, 1992; MENVIQ and MSSS, 1993.

Information Supplement

LEAD SHOT CONTAMINATION OF WATERFOWL ALONG THE ST. LAWRENCE

The ingestion of shot from hunting rifle cartridges is a major source of lead contamination in aquatic birds. A typical cartridge used in waterfowl hunting contains several hundred pellets of lead, most of which end up in the environment, especially in wetlands. Birds can be poisoned as a result of ingesting these pellets, which remain in their gizzards for up to one month. Following ingestion, lead accumulates rapidly in bony tissue and becomes stable. Lead poisoning, also called saturnism, can cause digestive, circulatory and nervous system disorders, and may even lead to death. Dabbling ducks (Mallard, Black duck, teal, etc.) are the most vulnerable to lead poisoning. Lead poisoning in waterfowl is likely to occur in areas of intense hunting activity because shot accumulates on the bottom of marshes and swamps there.

A study by the Ministère du Loisir, de la Chasse et de la Pêche (1987-1988) on the quantity of lead shot found in gizzards of waterfowl in Quebec showed that, in three out of ten regions, dabbling ducks (Mallard, Black duck, teal, etc.) had lead concentrations above 5%, meaning that the ducks may show symptoms of lead poisoning. Two of these regions – Trois-Rivières (8.4%) and Montreal (6.8%) – are near the St. Lawrence and are, moreover, characterized by the most intense waterfowl hunting.

The species with the highest rates of ingestion are also those that are the most heavily harvested (Mallard, Black duck and hybrids of these two species). Approximately 10.5% of the gizzards of Mallards, Black ducks and hybrids of the two species sampled in the Trois- Rivières region contained at least one lead pellet. In the Montreal region, 8.9% of the gizzards of this group of species contained at least one pellet; however, in this region, 47% of the ducks that had ingested lead shot were from Venise-en-Québec, near Lake Champlain, whereas in the Trois-Rivières region, most of the ducks were taken in a narrow strip of land along the St. Lawrence where hunting effort is greatest.

FIGURE 3.13.4

Rate of lead shot ingestion in dabbling ducks and diving ducks during the hunting season, by riverfront administrative region (1987-1988)

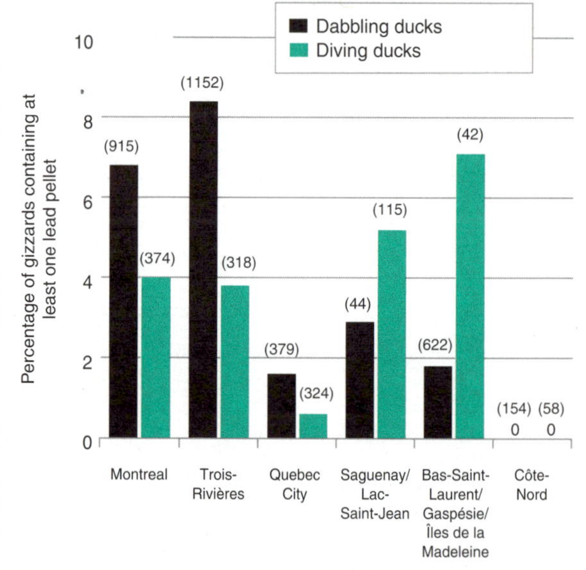

Note: Dabbling ducks considered here include: the Mallard, Black duck + Mallard, Black duck, Gadwall, American wigeon, Green-winged teal, Blue-winged teal, Northern shoveler, Northern pintail and Wood duck.

Diving ducks considered here include: the Redhead, Canvasback, Greater scaup, Ring-necked duck, Common goldeneye, Barrow's goldeneye, Bufflehead, Ruddy duck, Common merganser, Red-breasted merganser and Hooded merganser.

Figures in brackets indicate the number of samples.

Source: Based on data from Lemay et al., 1989.

In riparian areas, the percentage of diving duck gizzards (scaups, goldeneyes and mergansers) containing lead was 7.1% for the Bas-Saint-Laurent–Gaspésie–Îles de la Madeleine, 5.2% for the Saguenay–Lac-Saint-Jean region, 4% for Montreal, 3.8% for Trois-Rivières, 0.6% for Quebec City and 0% for the Côte-Nord. Diving ducks appear to be naturally resistant to lead poisoning, given that high lead concentrations in the blood or bones are not associated with weight loss or other symptoms of lead poisoning. In general, geese, Canada geese, and sea ducks contained very low lead levels in their tissues.

In a Canada-wide study done in 1988-1989, lead contamination in waterfowl was assessed by examining the wings of young ducks taken by hunters. In Quebec, 2408 Mallards and Black ducks (dabbling ducks) and 259 Ring-necked ducks (diving duck) were examined. High lead levels (10 mg/kg and over in bone) were found in 19% of the dabbling duck specimens (Mallards and Black ducks) and in 50% of the Ring-necked ducks. Sectors characterized by intense hunting (500 hunter-days and over) and high lead concentrations (10 mg/kg and over) were likely to exhibit lead poisoning problems. In Quebec, the areas are the following: an area near Gaspé, the southeastern tip of Lake Saint-Jean, Baie-Comeau, and the banks of the St. Lawrence from Île d'Orléans to the Ontario border.

Sources: Based on data from Lemay et al., 1989; Kennedy and Nadeau, 1993.

Information Supplement

THE RETURN OF THE ATLANTIC SALMON

In the St. Lawrence ecosystem, a number of fish stocks have been decimated, and many species are at risk in both the freshwater and saltwater sectors. Although it has experienced a decline of its own, it looks as though the Atlantic salmon is making a comeback.

The Atlantic salmon, prized for its flesh and high market value, as well as its large size and combative nature, has long been at the centre of a tug-of-war between commercial and sport fishermen. Competition between these two groups became increasingly fierce in the 1960s with the sharp decline in the stock. By 1972, the species' situation was so alarming that the Quebec government ordered a halt to all commercial fishing around the Gaspé peninsula, and three of the four Maritime provinces followed suit. The most credible hypothesis blamed the decline on excessive commercial fishing, but not necessarily in Canadian waters. Instead, our salmon were being caught while growing out in the icy waters off the west coast of Greenland. In the early 1970s, the international fleet operating on the high seas there was removing more than 2000 tonnes of Atlantic salmon annually, of which a large proportion originated from Quebec rivers. In 1982 and 1983, the commercial fishery was reopened on a trial basis, but it soon became clear that the resource was severely depleted. Commercial fishing was thus banned again, this time indefinitely, and the ban was extended to the north shore of the Lower Estuary and part of the upper north shore of the Gulf.

In 1983, all the countries sharing the resource negotiated an agreement which led to a drastic reduction in high seas fishing and the establishment of a modest quota for Greenland, the only country permitted to keep harvesting the species. As well, the North Atlantic Salmon Conservation Organization (NASCO) helped to impose a moratorium on the commercial fishery in Newfoundland and reduce fishing activity in Labrador. Fisheries and Oceans Canada set up a voluntary licence buy-out program and in 1992 half of the licence holders gave up their licences. Meanwhile, Quebec has gradually been buying back all the remaining commercial fishing permits on the Moyenne-Côte-Nord (between Sept-Îles and Natashquan), because it is believed that fishermen using traps along the shore and gillnets are intercepting too many of the salmon migrating toward rivers in Quebec, the Maritimes and the state of Maine.

Prospects for the future are good, however, especially for anglers, since the Quebec government has decided to allocate most of the salmon to them. Even if subsistence fishing, which is limited to Native peoples, continues to be the primary type of salmon fishing, it is anglers who will catch the most salmon. Analyses have suggested that this type of harvesting injects eight to ten times more money into the regional economy than the commercial fishery does. Moreover, there is almost no risk of overharvesting by anglers. With more than 115 salmon rivers and an annual catch of roughly 20 000 fish on 93 of these rivers, the sport fishery can generate major economic spinoffs for nearby regions. The current commercial catch, which stood at 19 363 in 1992, will eventually be transferred to the sport fishery, thereby doubling its net potential. New facilities on some rivers and judicious restocking will probably further enhance the sport fishery's potential.

Despite its complex life cycle, the Atlantic salmon may be one of the easiest species to manage.

Spawners can be counted singly and the economic benefits of exploiting the resource are distributed at the local level, thus encouraging its protection. Our present knowledge of the species' needs is satisfactory. Of course, considerable effort is needed to manage the resource and an enormous geographic area is involved; however, there is every indication that the Atlantic salmon will make a strong comeback.

Source: Based on data from Fisheries and Oceans, 1991; Caron et al., 1993.

SPORT HUNTING AND FISHING

Main Findings

- The average annual harvest of waterfowl along the St. Lawrence totalled more than 370 000 birds between 1988 and 1991, reflecting the widespread popularity of hunting.
- In the mid-1980s, pike, Yellow perch, Walleye and Atlantic tomcod were the most heavily fished freshwater species.
- Freshwater fish caught in the sport fishery are subject to consumption restrictions, which have generally been eased since 1985. In the Trois-Rivières–Quebec City sector, the guidelines for Walleye and Smallmouth bass were raised from a maximum of two meals per month to four and from two to eight for Yellow perch in Lake Saint-Louis between 1992 and 1993. Northern pike caught in Lake Saint-Louis and Walleye taken in Repentigny–Sorel were subject to the most stringent restrictions during this period.

Relative Importance

- Sport hunting and fishing rank fifth of the seven most-influenced characteristics for the St. Lawrence as a whole.

Recommendations on Monitoring

- Sport hunting: In conjunction with harvest levels – since they can fluctuate for a number of reasons (economic and climatic conditions, and the popularity of the activity) – an indicator should be developed that can monitor changes in hunting effort.
- Sport fishing: This activity is affected by prevailing economic conditions, among other factors. Catch rates could be supported by data on the effort required to meet the existing quota, according to the species considered.
- Restrictions on consumption: It is important to continue tracking changes in restrictions relative to contaminant levels in fish. A system should be put in place to evaluate the impact of these restrictions on fishermen's real consumption of fish, and on the health of fishermen who do not abide by the guidelines.

3.14

ACCESSIBILITY OF THE BANKS AND RIVER

Jean Burton

Background

In its broad sense, the concept of accessibility has numerous facets, such as conflicting uses, legislative measures, private ownership of land and infrastructures and the nature of the latter. All these factors may limit or promote physical access to the banks, whether for recreational and tourism purposes or residential and industrial development, among other desired uses. Unique or rare components, such as particular landscapes and resources especially prized by users, are included in the concept of accessibility (see information supplement entitled, "Whale watching: An industry to keep an eye on"). In addition, the quality of river resources to which users want access, whether they be biological, aesthetic or of some other nature, must be adequate for the activity concerned.

The two indicators selected to evaluate this characteristic – the number of public beaches open and the number of recreational and tourism infrastructures – measure only the extent to which the River and its banks are accessible for recreational and tourism purposes. For example, while a public beach offers physical access to the shoreline, the water there must meet certain quality criteria in order for swimming to be permitted. The number of infrastructures is an indication of the distribution and abundance of sites that provide access to the banks, but does not shed any light on their quality.

Interpretation

The 21 beaches in the public-beach network were all open in 1992; about 10 of them are located in the Montreal region. Bacterial water quality at the beaches has improved since 1990.

Most of the boat launching ramps, marinas and wharves are found along the main course of the River. The Lower Estuary and Gulf have the majority of rest areas, scenic lookouts and observation sites. Changes in this inventory since 1987 could not be assessed.

Although the available data are too incomplete to assess changes in the accessiblity of the banks and River, this characteristic may be influenced by certain factors or by indications (losses, restrictions, alterations) over the next few years: the frequency and extent of water quality criteria exceedances for primary-contact recreation, water quality in tributaries, surface area of protected sites, urban and industrial wastewater discharges, condition of biological resources and restrictions on consumption of fish.

The number of public beaches open

Definition

In 1992, the total number of swimming areas along the River was unknown; only the 21 public beaches subject to monitoring of bacterial water quality were officially identified. Sixteen public beaches are located in fresh water, and five in salt water. Many other sites along the St. Lawrence are used for swimming too, although there is no official monitoring of water quality.

Public beaches in fresh water are classified on the basis of the fecal coliform count per 100 mL; the number of enterococci per 100 mL is considered the best indicator of risk to human health in the marine sector. The water in swimming areas must not contain more than 200 fecal coliforms/100 mL, or 36 enterococci/100 mL.

On the basis of this measurement system, beaches may receive one of four possible ratings: A or excellent (0 to 20 fecal coliforms or 0 to 5 enterococci per 100 mL); B or good (21 to 100 fecal coliforms or 6 to 20 enterococci per 100 mL); C or fair (101 to 200 fecal coliforms or 21 to 35 enterococci per 100 mL); D or polluted (over 200 fecal coliforms or more than 36 enterococci per 100 mL).

Limitations

Annual sampling of water in swimming areas is conducted only at public beaches that meet the set safety criteria of the Ministère du Travail du Québec, under the terms of an agreement in principle with the Ministère de l'Environnement et de la Faune. Participation in the *Programme Environnement-Plage* (beach water quality monitoring program) is voluntary, but entails mandatory application of Ministère du Travail standards. Hence, beach status is restrictive. The granting and withdrawal of beach operating permits is linked to a region's ability to draw tourists and/or the permit holders' financial situation. Beach openings and closings during the season and from year to year may be linked not only to water quality factors but also to socio-economic considerations.

PROBLEM OR RISK-PRONE SECTORS

FLUVIAL SECTION | FLUVIAL ESTUARY
UPPER ESTUARY AND SAGUENAY | LOWER ESTUARY AND GULF

Judging from the most recent data, none of the sectors is problematic. However, on the basis of water quality at previously monitored beaches, the Fluvial Section (Lake Saint-Louis, lesser La Prairie basin and Lake Saint-Pierre), the Saguenay River, the Upper Estuary and Chaleur Bay are all at risk.

Results

In 1992, bacterial water quality at the 21 public beaches along the River, Lake des Deux Montagnes and Saguenay River ranged from excellent (A) to good (B), except at one beach, which received a rating of fair (C). All these beaches remained open between 1989 and 1992. Bacterial water quality at beaches in the Lake Saint-François (Nos. 1 to 4) and Lake des Deux Montagnes (Nos. 5 to 13) sectors improved markedly (see Figure 3.14.1). Beach 14, on the Saguenay, had the highest fecal coliform count of all 21 public beaches during the three years from 1990 to 1992.

In 1989, 9 of the 13 freshwater beaches on the Fluvial Section received an A rating, three were rated B and one was not rated. In 1990, water quality at these beaches deteriorated slightly, given that four were rated A, seven were rated B and two were not rated. In general, water quality has improved appreciably since 1990, as most of the beaches (11 out of 13 in 1991 and 12 out of 13 in 1992) received an A rating. Of the three beaches on the Saguenay River (Nos. 14 to 16), two received an A rating between 1989 and 1992. At beach 14, water quality deteriorated between 1989 and 1991, dropping from A in 1989 to C in 1991; however, in 1992, a substantial improvement was noted (B rating).

FIGURE 3.14.1

Year-to-year change in water quality at 21 public beaches (1989-1992)

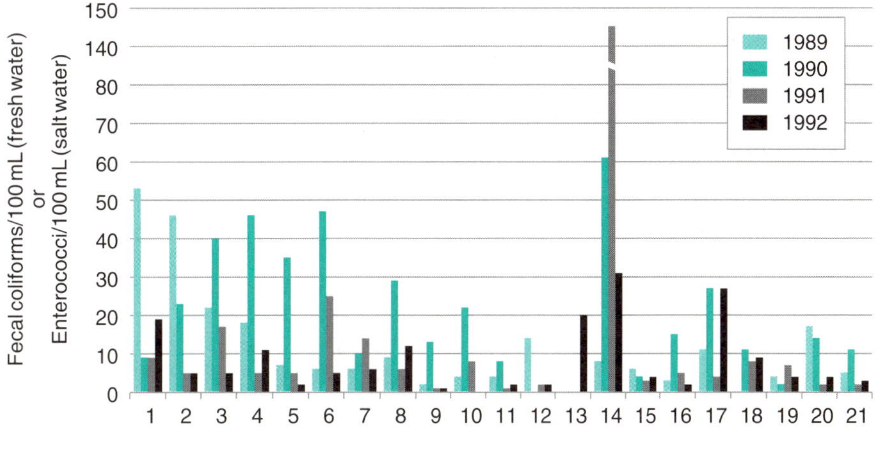

1 Baie du Village
2 Camp Mont-Immaculé
3 Municipal beach
4 Parc régional des Îles
5 Municipal beach
6 Notre-Dame-de-Fatima
7 Sainte-Madeleine-de-Rigaud
8 Paul Sauvé Park, Oka
9 Camp Notre-Dame, Oka
10 Plage Roger Inc.
11 Cap Saint-Jacques (site 1)
12 Cap Saint-Jacques (site 2)
13 Pointe aux Carrières
14 Club Voile Saguenay
15 Camping Dam-de-Terre
16 Colonie Notre-Dame
17 Saint-Siméon
18 Community beach
19 Haldimand beach
20 Penouille beach
21 Beau Bassin beach

Note: Where no bar appears, the data were unavailable.

Source: Based on data from Léveillé, 1992.

In 1989, two of the five beaches located in the salt water part of the River received an A rating, two a B and one was not rated. In 1990, as in the case of fresh water, a slight deterioration was noted given that only one beach had an A rating, three had a B and one a C. Since 1990, water quality has improved at four beaches out of five. At beach 17 in the Upper Estuary, water quality improved considerably in 1991 (A rating), but deteriorated again in 1992 (C rating).

Between 1987 and 1990, 36 other beaches were monitored under the provincial program, but ceased to be included in the program as of 1992. Nine were rated as polluted (D) during their last year of operation. Three of these closed beaches were in the freshwater sector (on Lake Saint-Pierre), and the other six in the saltwater sector (one in the Upper Estuary, four along the Gaspé coast, including, three in Chaleur Bay, and one in the Îles de la Madeleine).

The number of recreational and tourism infrastructures: Boat launching ramps, marinas, wharves, rest areas, scenic lookouts and observation sites

Definition

Various infrastructures allowing access to the River and banks (boat launching ramps, marinas, wharves, rest areas, scenic lookouts and observation sites) were inventoried during the period 1987 to 1991, and the data are presented by hydrographic region.

This indicator can illustrate changes in access to the River and specific aspects of the recreational and tourism activities in the regions. It can also provide a picture of environmental stress.

Along the St. Lawrence, 456 infrastructure elements providing access to the River were inventoried: 91 marinas, 93 wharves, 57 rest areas, 26 scenic lookouts and 26 observation sites. The pleasure-boating infrastructures (boat launching ramps, marinas and wharves) appear to be concentrated in the main course of the River. Infrastructures used for landscape and nature watching (rest areas, observation sites and scenic lookouts) are located mainly upstream from Quebec City, in the Côte-Nord, Bas-Saint-Laurent, Gaspé and Îles-de-la-Madeleine.

Limitations

The data are derived from a variety of sources, and changes in the inventories since 1987 cannot be determined. The indicator does not take other types of infrastructure providing access to river resources, such as roads and residential developments, into account.

Problem or Risk-Prone Sectors

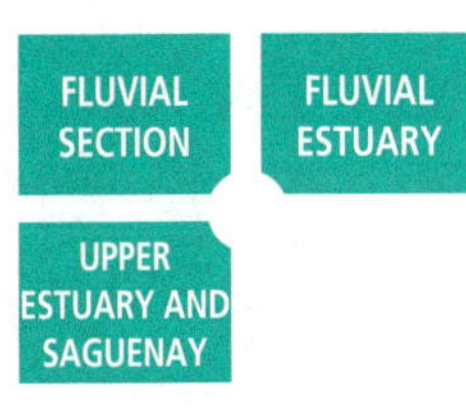

Varies according to the type of infrastructure.

Results

Based on the infrastructures inventoried between 1987 and 1991, two different developmental thrusts can be seen. The Fluvial Section contains the largest number of boat launching ramps (58%), marinas (56%) and wharves (43%) (see Figure 3.14.2). By contrast, rest areas, scenic lookouts and observation sites are more numerous in the Lower Estuary and Gulf (47%, 37% and 63%, respectively). However, these three types of infrastructure were not included in the inventory of the Saguenay River in SLC, 1993b.

While the main course of the River also has 19% of the rest areas, no scenic lookouts or observation sites were listed for this sector. The Lower Estuary/Gulf sector has 24% of the wharves, 20% of the marinas and 18% of the boat launching ramps.

The Fluvial Estuary has one of the lowest proportions of boat launching ramps (13%), marinas (11%) and wharves (17%). In contrast, 19% of the rest areas, 39% of the scenic lookouts and 23% of the observation sites are located there. This part of the River is especially narrow and its scarped banks are suitable for lookouts offering scenic views of the River. Whereas most of the lookouts are on the north shore of the Fluvial Estuary, the majority of rest areas are on the south shore.

FIGURE 3.14.2

Inventory of various infrastructures accessing the River (1987-1991)

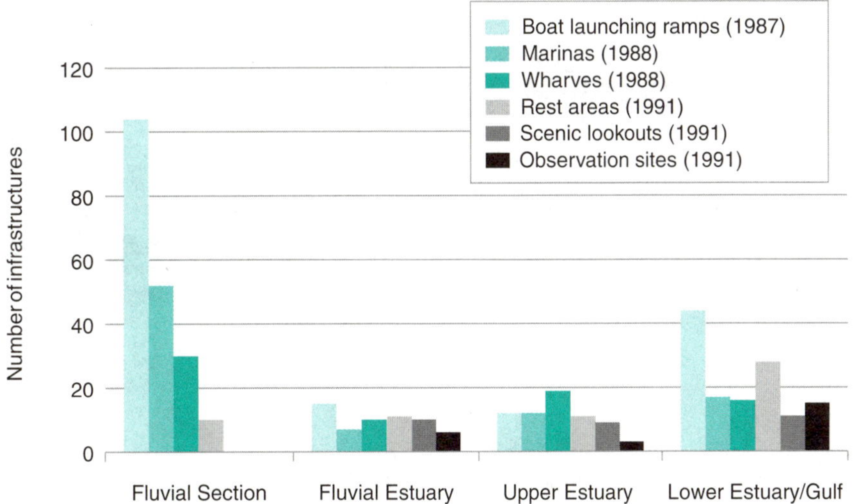

Note: Where no bar appears, the data were unavailable.

Sources: Based on data from MLCP, 1987, 1982; Canadian Coast Guard, 1989; Ministère du Tourisme du Québec and regional tourism associations, 1991; MTQ, 1991.

In the Upper Estuary, the proportion of boat launching ramps (11%), marinas (13%), wharves (16%) and rest areas (11%) is also very small. This sector contains 19% of the scenic lookouts and 19% of the observation sites.

Information Supplement

WHALE WATCHING: AN INDUSTRY TO KEEP AN EYE ON

The Lower Estuary of the St. Lawrence is one of the best places in the world to see whales. Many species of baleen whales cross the Gulf and enter the estuary between June and October. Whales can be seen spouting and rolling in the glacial waters there, which teem with crustaceans and small pelagic fish, a key food source for them. From land-based sites and vessels of all sizes, tourists can now observe Blue whales, Humpback whales, Fin whales and Minke whales. During the whale-watching season, excursion boats set out from the ports of Rivière-Portneuf, Sault-au-Mouton, Les Escoumins, Grandes-Bergeronnes, Tadoussac and Baie-Sainte-Catherine. Cruises also depart from a few south shore ports like Trois-Pistoles and Rivière-du-Loup.

The presence of belugas in the River and the appearance of great whales give a considerable boost to marine and economic activity. The whale-watching industry drew more than 120 000 people in 1991 (see Figure 3.14.3). In 1992, the industry generated direct and indirect spinoffs which likely exceeded $15 million. Nine companies operated 15 whale-watching cruise vessels in 1991. With a daily capacity of over 3000 passengers and a season lasting about 100 days, the industry could undergo considerable expansion without any increase in its current equipment base. Despite a slowdown attributed to two disastrous years in a row (1991 and 1992), the industry does not appear to have reached the saturation point. Nonetheless, the ever-greater number of boats navigating around marine mammals is a matter for concern; the impact that whale-watching vessels (and the growing number of boats in the region) may have on the equilibrium of the environment is not really known. As such, these vessels are subject to regulations designed to eliminate or reduce the risk of harassment or disturbance to belugas and other whales.

Whales can also be observed from land-based sites in Forillon National Park, and boat cruises are available in the Mingan area. The Mingan marine excursions are unusual in that scientists take part, using a photographic technique to identify marine mammals and track their movements between the Lower Estuary and the Gulf, and between the Gulf and their overwintering areas in the West Indies.

FIGURE 3.14.3
Number of cruise boat passengers (1984-1991)

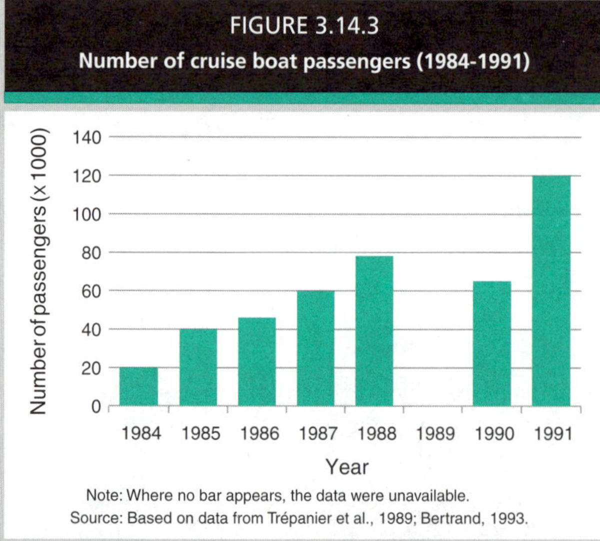

Note: Where no bar appears, the data were unavailable.
Source: Based on data from Trépanier et al., 1989; Bertrand, 1993.

Both in the Lower Estuary and in the Mingan area, a wide variety of marine mammals such as dolphins, porpoises, Sperm whales and Killer whales can be found alongside rorquals, as well as at least two seal species – the Harbour seal and the Grey seal.

Source: Based on data from Trépanier et al., 1989; Bertrand, 1993.

ACCESSIBILITY OF THE BANKS AND RIVER

Main Findings

- Bacterial water quality at the 21 monitored public beaches has improved since 1991, making the River more accessible for recreational and tourism activities. Ten public beaches in the Lake Saint-François and Lake des Deux Montagnes sectors registered the largest reductions in fecal coliform counts between 1990 and 1992.
- Water quality measurements for primary-contact recreation indicate that there may be some problem or risk-prone areas in both upstream sectors; certain species of game fish are at risk and restrictions on consumption of freshwater fish remain in place; the Beluga whale is still endangered, thus the need for controls on whale watching activities; owing to bacterial water quality in shellfish areas, shellfish harvesting is restricted in certain parts of the marine sector. All these situations are warning signs that some recreational and tourism uses of the River may be lost in the future.
- Any deterioration in the current signs of environmental disturbance could affect access to the River and its banks over the coming years. Signs of disturbance are important since they can give rise to restrictions on or changes in access to the River and its banks, whatever the desired use.

Relative Importance

- Accessibility of the River and its banks is the second of the seven most-influenced factors related to the St. Lawrence as a whole.

Recommendations on Monitoring

- The indicators used relate solely to access to the River and its banks for recreational and tourism purposes. In this sense, the status of public beach is of limited use, given that whether or not beaches are open during the season or year may depend on water quality factors but also on socio-economic considerations. To improve this indicator, all swimming areas accessing the River should be inventoried, and water quality at representative samples of them should be tested annually.

- An inventory of the infrastructures should be carried out periodically, by hydrographic region, to track any changes. In addition, the relationships between the number of infrastructures and their rate of use along with the population density in the river sector where they are located should be determined in order to evaluate the pressures on access to the shoreline areas in question.
- An indicator needs to be developed that measures other aspects and incorporates physical access in relation to the quality of resources sought by users. The indicator should be weighted according to the specific context of the different hydrographic regions.

Summary and Conclusions

This assessment of the state of the St. Lawrence River has produced conclusions that are useful to improve decision making and subsequent environmental monitoring. When the analysis was completed, a brief evaluation of the selected characteristics was drawn up to show whether their condition had improved, stabilized or deteriorated. If there was insufficient data to determine the state of a characteristic, it was considered undetermined. This information is presented in Table 4.1.

Overall, six characteristics that are indicative of the state of the St. Lawrence have improved since the late 1970s: sediment quality, urban wastewater discharges, industrial wastewater discharges, modification of the floor and of hydrodynamics, water quality in the St. Lawrence, and natural environments and protected species. Only one characteristic, commercial fishing, has deteriorated since this period. Three other characteristics that have not improved or deteriorated since the late 1970s were considered stable: water quality in the tributaries, sport hunting and fishing, and shipping. There was insufficient data to evaluate four other characteristics: condition of biological resources, accessibility of the banks and River, biodiversity and shoreline modifications.

The main causes of the ecological imbalance of the River are toxic substances and bacteria, some of which were identified more than 15 years ago. Water quality in the tributaries was also identified at that time, and it is still a major factor in the deterioration of water quality in the St. Lawrence. Organic contaminants used in agriculture are particularly detrimental. Effluents discharged by certain industries contain toxic substances and the total toxic load added by all riverside industries has still not been quantified.

LARGE AMOUNTS OF TOXIC SUBSTANCES

Toxic substances in the River come from industrial, urban and agricultural activities. The Great Lakes–St. Lawrence system drains one of the most highly industrialized regions in North America. Farming is a major activity in both upstream sections of the St. Lawrence and a variety of contaminants are discharged into the River by its 244 tributaries. Agriculture and atmospheric inputs from local and remote emissions are significant nonpoint sources of pollution.

TABLE 4.1

Evaluation of selected characteristics of the St. Lawrence and changes in them since the late 1970s

Characteristic	Evaluation	Explanation
1. Sediment quality	*Improved*	Contamination levels decreased, but sediments are still contaminated.
2. St. Lawrence water quality	*Improved*	Discharges of toxic and bacterial contamination decreased, but restrictions on use are still in force.
3. Tributary water quality	*Stable*	Urban, industrial and agricultural inputs continue to contribute to chemical and bacterial contamination.
4. Biodiversity	*Undetermined*	Although the number of species at risk seems to be increasing, lack of knowledge prevents an assessment.
5. Natural environments and protected species	*Improved*	Protected areas and species increased.
6. Condition of biological resources (abundance and contamination)	*Undetermined*	Too many species-to-species fluctuations were found to make an evaluation.
7. Shipping	*Stable*	The number of ships remained constant and risks have not increased.
8. Modification of the floor and of hydrodynamics	*Improved*	The floor of the River was modified in specific areas only.
9. Shoreline modifications	*Undetermined*	No data available on loss of wetlands since the late 1970s.
10. Urban wastewater discharges	*Improved*	Contamination by organic and inorganic substances and bacteria was reduced.
11. Industrial wastewater discharges	*Improved*	Discharges of liquid toxic wastes decreased.
12. Commercial fishing	*Deteriorated*	Certain fish stocks collapsed.
13. Sport hunting and fishing	*Stable*	Contamination of species fished means restrictions on consumption are still necessary. Waterfowl harvest remained steady or increased slightly.
14. Accessibility of the banks and River	*Undetermined*	Recreational and tourist activities increased; however other facets of accessibility were not considered.

Industrial wastewater discharges

At present, data is available only on the 50 industrial plants targeted under the St. Lawrence Action Plan. Of these, 24 are located between Cornwall and Quebec City and discharge their effluent directly into the St. Lawrence, accounting in 1991 for about 9% of its toxic load. Another 6300 smaller plants are located along the St. Lawrence; 4000 of them are located in the MUC. Because of hydrodynamic features, discharges remain near the shore, and have a direct impact on bioresources and uses of the River. The 50 targeted plants are in the pulp and paper, metallurgy, organic chemicals and inorganic chemicals sectors. The greatest toxic inputs are received by two upstream sectors, followed by the upper parts of the Upper Estuary and Saguenay, and the Baie-Comeau area in the Lower Estuary.

In 1992, the volume of process wastewater discharged by these industries was over 1.4 million cubic metres per day (56% from pulp and paper mills and 35% from metallurgical plants). These discharges contained 272 000 kg/day of suspended solids. In 1992, industrial effluents still contained many toxic substances (the Chimiotox index measures 124 substances), including oils and greases, heavy metals and other metals (such as aluminum, iron and manganese), PCBs, resin acids and fatty acids. No data is available on discharges and pH values of dissolved or suspended solids.

Potential Ecotoxic Effects Probe (PEEP) measurements of organisms exposed to an industrial effluent and not in a natural environment show, for the River as a whole, higher potential toxicity indexes for the pulp and paper and inorganic chemistry sectors.

Water quality in the tributaries

The waters of the tributaries are loaded with a variety of toxic substances from industrial and agricultural activities and urban wastewater discharges. The contribution of 51 tributaries to the toxic load in the St. Lawrence River was measured at their mouths in 1991 for 19 contaminants. An estimated 29% of the toxic load in the St. Lawrence comes from these tributaries between Cornwall and Quebec City. The tributaries with the largest mean annual discharges also contribute the largest total toxic loads (organic and inorganic). Among the tributaries with loads of more than 100 000 toxic units are the Saint-Maurice, Ottawa, Richelieu, Saint-François, Batiscan, Yamaska and Chaudière rivers. It is impossible to tell whether toxic loads have decreased or increased since 1978 for lack of adequate data.

About 20 of the 51 tributaries contribute 97% of the total toxic load discharged to the St. Lawrence. Organics constitute 96% of this toxic load, testifying to the contribution of agricultural activities to contamination of the St. Lawrence.

Urban wastewater discharges

The contribution to the toxic load from treated and untreated urban wastewater discharges cannot be evaluated at present. Toxic substances include oils and greases, solvents, domestic products and effluent discharges from small industries.

WIDESPREAD PRESENCE OF PATHOGENIC BACTERIA (FECAL COLIFORMS)

Urban wastewater discharges and agriculture are the main sources of bacterial contamination; it is not possible to evaluate the relative importance of these two sources using the available data. In 1992, 1.35 million inhabitants, 33% of riverside populations, continued to discharge their untreated wastewater into the St. Lawrence. Treated urban wastewater containing dissolved organic matter and bacteria and discharged through outfalls also contributes to contamination. There is no existing measurement of discharges from tributaries, but water quality indexes for the period 1990-1993 registered coliforms in all tributary water masses examined.

ADDED PRESSURE FROM FISHING ON FISH SPECIES IN SERIOUS DECLINE AND AT RISK

Rarely can the specific cause or causes of the destabilization of the stock of a particular species be pinpointed; a whole series of factors is generally responsible. In the case of migrating species, the causes may even lie beyond our borders. The collapse of a stock can have major consequences on commercial and sport fishing, as shown by the collapse of the Atlantic salmon fishery. Atlantic salmon have been decimated for over 10 years, and the slow recovery of the stock will limit salmon exploitation to sport fishing for some time to come.

The serious decline in abundance observed between 1986 and 1992 in certain freshwater and saltwater fish stocks or species (some of which are at risk) was accompanied by relatively heavy commercial fishing in the same period, particularly of Rainbow smelt, American eel, Atlantic tomcod and Atlantic cod. Moreover, complementary information on commercial landings of Atlantic sturgeon shows that catches consisted solely of fish that had not yet reached sexual maturity.

INSUFFICIENT PROTECTED NATURAL ENVIRONMENTS

The survival of many resources that are typical of the St. Lawrence may depend on the protection of natural environments. Analysis of 1992 data on the River shows an inequity between the location of protected areas and the distribution of animal and plant species at risk. The most threatened animal species and communities are in the two upriver sectors. The largest number of spawning grounds are in Lake Saint-Pierre and the Sorel islands. In terms of the distribution of rare and threatened plant species and communities along the St. Lawrence, it is clear that the main course of the River and the Fluvial Estuary have the greatest need for protected areas. Yet there are relatively few protected natural environments in the Fluvial Section (20 653 ha), and they are almost nonexistent (886 ha) in the Fluvial Estuary. There are still unprotected natural environments with good potential in the littoral fringe of this sector (such as from Saint-Augustin-de-Desmaures to Sainte-Anne-de-la-Pérade). But the surface area of protected spaces is only one element to be considered. Surface area may vary depending on the nature of the space protected and its role in the life cycles of different species.

Analysis shows that, in addition to the four main causes of disturbances, the fluvial ecosystem is subject to other risks. The risks are related to three characteristics which have no great influence on the River as a whole but which do affect specific parts of it. The three characteristics are shoreline modifications, shipping, and modification of the floor and of hydrodynamics.

ADDITIONAL WETLAND LOSSES

The St. Lawrence lost considerable wetland areas between 1945 and 1976. It is impossible to evaluate wetland losses or gains since 1976 because the only inventory available is for the 1990-1991 period and covers only the River corridor. Given past losses and the acknowledged ecological and economic importance of wetlands, additional loss of wetlands could have serious repercussions on the condition of the river ecosystem, entailing further losses of shoreline habitats essential to wildlife, the disappearance of plant species, reduction of the River's capacity for self-purification, and loss of biological diversity.

RISKS ASSOCIATED WITH SHIPPING: ACCIDENTAL SPILLS AND INTRODUCTION OF EXOTIC SPECIES

In 1992, 17.4 million tonnes of dangerous goods (out of a total of some 95 million t) were handled in the commercial ports on the St. Lawrence; almost 90% of the tonnage was handled in the main course of the River or in the Fluvial Estuary. In 1992, 48 tanker-trips (out of a total of 880) in the North Traverse section of the ship channel had a draft that exceeded the minimum water depth. Since 1988, the percentage of trips where ship's draft exceeded this level has varied between 3.36% and 5.57%. In 1993, 75 accidental spills from ships and 78 of unknown or terrestrial origin were recorded; ships spilled an average of 475 gallons of oil.

Exotic species can encroach upon native species, disturbing food webs and competitive relationships among living communities. The effects of Zebra mussel colonization in the River corridor and the more recent introduction of the Quagga mussel are not well known. These two species colonized the St. Lawrence after being introduced into the Great Lakes through the release of ballast water from commercial ships.

PHYSICAL CHANGES TO HABITATS

The maintenance dredging of ports, the St. Lawrence Seaway and the ship channel alters the river floor and changes the flow of the River; between 400 000 to 600 000 m^3 of sediment has been dredged annually since 1989. The largest volumes are removed in the Fluvial Estuary and Lower Estuary; in 1983 and 1988, volumes were particularly high. The repercussions of dredging on habitats, water quality and the condition of biological resources are poorly understood. Dredging physically alters habitats and may resuspend contaminated sediments that could be taken up by various organisms.

The construction of shoreline infrastructures to provide access to the River for recreational and harbour activities also alters habitats physically, as does the filling and clearing involved in road building and urban expansion along the shore.

In light of these findings, it appears essential that systematic environmental monitoring of the St. Lawrence be undertaken to obtain a more accurate picture of the condition of the River and how it is changing.

In future years, it will be possible to compare data on the changing conditions of the St. Lawrence. But these data must be improved, enhanced and refined. Studies have too often been disconnected in space and time, and this has greatly limited their interpretation. Some data were over 10 years old; others covered only one part of the River. Sometimes data were available for only one year, becoming a point of reference for future assessments. The dynamic relationships of the river ecosystem make for complex study. The vastness of the River, the specific aspects of each hydrographic region and the many uses associated with human activities further complicate research.

The key element in the process was to consider the river ecosystem as a whole while showing the links among its components. The challenge that lies ahead will be to use environmental information to promote sustainable development on many levels – locally, riverwide, and even globally.

REFERENCES

Only documents used in the preparation of this volume are shown. For more information, readers may consult the detailed references in parts 1, 2, 3 and 4 of Volume 1: *The St. Lawrence Ecosystem.*

AMÉNATECH. 1992a. *Cartographie des marais, marécages et herbiers de Cornwall à Trois-Rivières pour 1 km de rive avec le capteur MEIS-II.* Prepared for Environment Canada, Conservation and Protection, Quebec Region, St. Lawrence Centre, Montreal.

AMÉNATECH. 1992b. *Cartographie des marais, marécages et herbiers le long du fleuve Saint-Laurent de Trois-Rivières à Montmagny au moyen de la télédétection aéroportée.* Prepared for Environment Canada, Conservation and Protection, Quebec Region, St. Lawrence Centre, Montreal.

AMÉNATECH. 1991. *Cartographie des milieux humides du Saint-Laurent avec le capteur MEIS-II, secteurs choisis entre Cornwall et Trois-Rivières.* Prepared for Environment Canada, Conservation and Protection, Quebec Region, St. Lawrence Centre, Montreal.

BÉLAND, P., S. DE GUISE, and R. PLANTE. 1992. *Toxicologie et pathologie des mammifères marins du Saint-Laurent.* St. Lawrence National Institute of Ecotoxicology, Rimouski.

BERTRAND, P. 1993. Personal communication. Environment Canada, Canadian Parks Service, Policy and Research Section, Québec.

BOUCHARD, I. 1993. *Bilan de la réduction des rejets des 50 industries du Plan d'action Saint-Laurent. Phase I.* Environment Canada and Ministère de l'Environnement du Québec, St. Lawrence Action Team, Montreal.

BOUCHER, P.R. 1992. Les milieux naturels protégés au Québec. Paper presented at the 21st congress of the Canadian Nature Federation, Québec, August 14, 1992. Ministère de l'Environnement du Québec, Direction de la conservation et du patrimoine écologique, Québec.

BSQ – BUREAU DE LA STATISTIQUE DU QUÉBEC. 1990. *La pêche maritime au Québec, 1956-1985. Statistiques économiques (données antérieures à 1984).* Direction des statistiques sur les industries, Québec.

BUSBY, D. 1993. Unpublished data. Environment Canada, Conservation and Protection, Atlantic Region, Canadian Wildlife Service, Sackville, New Brunswick.

CANADIAN COAST GUARD. 1994. Unpublished data of the Alert Network. Transport Canada, Laurentian Region.

CANADIAN COAST GUARD. 1993. Unpublished data on annual traffic in ports and various sectors of the river, 1980, 1984, 1988, 1989, 1990, 1991 and 1992. Transport Canada.

CANADIAN COAST GUARD. 1991. *Havres et ports. Harbours and Ports.* Transport Canada, Laurentian Region, Harbours and Ports Branch, Québec.

CANADIAN COAST GUARD. 1989. *Analyse des données recueillies lors du recensement 1988.* Transport Canada.

CARIGNAN, R., S. LORRAIN, and K.R. LUM. 1993. Sediment dynamics in the fluvial lakes of the St. Lawrence River: Accumulation rates and residence time of mobile sediments. Submitted to *Geochimica* et *Cosmochimica Acta.*

CARON, F., M. SHIELDS, and D. FOURNIER. 1993. *Bilan de l'exploitation du saumon au Québec en 1992.* Ministère du Loisir, de la Chasse et de la Pêche, Direction de la faune et des habitats, Service de la faune aquatique.

CASTONGUAY, M., P.V. HODSON, C.-M. COUILLARD, M.J. ECKERSLEY, J.D. DUTIL, and G. VERREAULT. 1994a. Why is recruitment of the American eel *Anguilla rostrata* declining in the St. Lawrence River and Gulf? *Canadian Journal of Fisheries and Aquatic Sciences* 51:479-488.

CASTONGUAY, M. P.V. HODSON, C. MORIARTY, K.F. DRINKWATER, and B.M. JESSOP. 1994b. Is there a role of ocean environment in American and European eel decline? *Fisheries Oceanography* 3:197-203.

CHAPDELAINE, G. 1993. Personal communication. Environment Canada, Conservation and Protection, Quebec Region, Canadian Wildlife Service, Québec.

CHAPDELAINE, G., P. LAPORTE, and D.N. NETTLESHIP. 1987. Population, productivity and DDT contamination trends of Northern gannets (*Sula bassanus*) at Bonaventure Island, Québec, 1967-1984. *Canadian Journal of Zoology* 65:2922-2926.

CHIASSON-DUFOUR, M. 1994. Personal communication. Fisheries and Oceans, Assets Management Branch, Québec.

CHOUINARD, G. 1994. Personal communication. Fisheries and Oceans, Sciences Branch, Groundfish Section, Moncton, New Brunswick.

COMITÉ D'ÉTUDE SUR LE FLEUVE SAINT-LAURENT. 1978. *Étude sur le fleuve Saint-Laurent.* Final report. Québec.

COSEWIC – COMMITTEE ON THE STATUS OF ENDANGERED WILDLIFE IN CANADA. 1993. *Species at Risk – List of Species with Designated Status.*

COSTAN, G., and N. BERMINGHAM. 1993. *Barème d'effets écotoxiques potentiels (BEEP): Une mesure synthétique du danger associé aux effluents industriels.* Environment Canada, Conservation and Protection, Quebec Region, St. Lawrence Centre, Montreal.

CWS – CANADIAN WILDLIFE SERVICE. 1990. *Répertoire des refuges d'oiseaux migrateurs et des aires de repos du Québec.* Environment Canada, Conservation and Protection, Quebec Region, Québec.

DALLAIRE, J.-P. 1993. *État des principales pêcheries de homard du Québec; les îles de la Madeleine et la Gaspésie.* Document de recherche sur les pêches de l'Atlantique No. 93/35. Canadian Atlantic Fisheries Scientific Advisory Committee (CAFSAC).

D'ANGLEJAN, B.-F., and M. BRISEBOIS. 1978. Recent sediments of the St. Lawrence Middle Estuary. *Journal of Sedimentary Petrology* 48(3): 951-964.

DE LADURANTAYE, R., C. DESJARDINS, R. NADEAU, Y. VIGNEAULT, and J.-F. LARUE. 1990. Les contaminants dans le Saint-Laurent. Bilan des connaissances dans le système aquatique marin. In *Symposium on the St. Lawrence: A River to be Reclaimed.* D. MESSIER, P. LEGENDRE and C.E. DELISLE, eds. Conference proceedings November 3-5, 1989, Montreal, Association des Biologistes du Québec. Université de Montréal, Environnement et Géologie series. Vol. II, pp. 105-134.

DE REPENTIGNY, L.-G. 1990. *Registre national des aires écologiques au Canada – Réserves nationales de faune au Québec.* Canadian Wildlife Service for Environment Canada, Canadian Council on Ecological Areas, Ottawa.

DRYADE (Le Groupe). 1981. *Analyse des pertes de végétation riveraine le long du Saint-Laurent de Cornwall à Matane (1945-1976).* Report No. 3683. Prepared for Environment Canada, Conservation and Protection, Quebec Region, Canadian Wildlife Service, Québec.

DUCHARME, J.-L., G. GERMAIN, and J. TALBOT. 1992. *Bilan de la faune 1992.* Ministère du Loisir, de la Chasse et de la Pêche, Direction générale de la ressource faunique, Québec.

ECKERSLEY, M.J. 1982. Operation of the eel ladder at the Moses Saunders generating station, Cornwall 1974-1979. In *Proceedings of the 1980 North American Eel Conference.* K.H. LOFTUS, ed. pp. 4-7. Ontario Fisheries Technical Report Series 4.

ENVIRONMENTAL PROTECTION BRANCH. 1992. Unpublished data of the Shellfish Water Quality Protection Program. Environment Canada, Conservation and Protection, Quebec Region, Montreal.

FERLAND, G. 1993. Personal communication. Port of Montreal.

FISHERIES AND OCEANS. 1993a. Unpublished data. Statistics and Informatics Division.

FISHERIES AND OCEANS. 1993b. *Marine Fisheries in Quebec – Annual Statistical Review 1991-1992.* Economics, Statistics and Informatics Branch, Statistics and Informatics Division, Québec.

FISHERIES AND OCEANS. 1992a. *Marine Fisheries in Quebec – Annual Statistical Review 1990-1991.* Economics, Statistics and Informatics Branch, Statistics and Informatics Division, Québec.

FISHERIES AND OCEANS. 1992b. *Table des marées et courants du Canada 1993. Volume 3, Fleuve Saint-Laurent et rivière Saguenay.* Canadian Hydrographic Service, Scientific Information and Publications Branch, Ottawa.

FISHERIES AND OCEANS. 1991. *Espèces en difficulté dans le Saint-Laurent.* "Du lac Saint-Pierre à Sept-Îles-Sainte-Anne-des-Monts." Series of 12 fact sheets.

FISHERIES AND OCEANS. 1990. *Marine Fisheries in Quebec – Annual Statistical Review 1988-1989.* Economics, Statistics and Informatics Branch, Statistics and Informatics Division, Québec.

FISHERIES AND OCEANS. 1988. *Marine Fisheries in Quebec – Annual Statistical Review 1984-1987.* Economics, Statistics and Informatics Branch, Statistics and Informatics Division, Québec.

FRÉCHET, A. 1994. Personal communication. Fisheries and Oceans, Maurice-Lamontagne Institute, Groundfish Section, Mont-Joli, Quebec.

FRENETTE, M., C. BARBEAU, and J.-L. VERRETTE. 1989. *Aspects quantitatifs, dynamiques et qualitatifs des sédiments du Saint-Laurent.* Projet de mise en valeur du Saint-Laurent. Hydrotech Inc. Consultants, for Environment Canada and the Government of Québec.

GHANIMÉ, L., J.-L. DESGRANGES, S. LORANGER, et al. 1990. *Les régions biogéographiques du Saint-Laurent.* Final report. Lavalin Environnement Inc. for Environment Canada and Fisheries and Oceans, Quebec Region.

GILLES SHOONER and Associates Inc. 1991. *Localisation des sites de reproduction des principales espèces de poisson du fleuve Saint-Laurent (Cornwall-Montmagny).* Atlas prepared for Fisheries and Oceans and Environment Canada, St. Lawrence Centre.

GRATTON, L., and C. DUBREUIL. 1990. *Portrait de la végétation et de la flore du Saint-Laurent.* Ministère de l'Environnement du Québec, Direction de la conservation et du patrimoine écologique, Québec.

GROUPE DE TRAVAIL SUR LES ESPÈCES DE FAUNE ET DE FLORE PRIORITAIRES DU COULOIR DU SAINT-LAURENT. 1993. Report prepared for the St. Lawrence Action Plan.

HAMMILL, M.O. 1995. Personal communication. Fisheries and Oceans, Maurice-Lamontagne Institute, Mont-Joli.

HARDY, B., J. BUREAU, L. CHAMPOUX, and H. SLOTERDIJK. 1991. *Caractérisation des sédiments de fond du Petit bassin de La Prairie, fleuve Saint-Laurent.* Environment Canada, Conservation and Protection, Quebec Region, St. Lawrence Centre, Montreal.

HENDRICK, A. 1991. "1990 operation of the eel ladder at the R.H. Saunders Generating Station, Cornwall, Ontario". *1991 Annual Report,* pp. 4-1 to 4-2. St. Lawrence River Subcommittee to the Lake Ontario Committee of the Great Lakes Fisheries Commission, Ontario Ministry of Natural Resources, Toronto.

HEUSMANN, H.W. 1974. Mallard/Black duck relationships in the northeast. *Wildlife Society Bulletin* 2(4): 171-177.

HODSON, P.V., C. DESJARDINS, É. PELLETIER, M. CASTONGUAY, R. McLEOD, and C.-M. COUILLARD. 1992. *Decrease in chemical contamination of American eels (*Anguilla rostrata*) captured in the estuary of the St. Lawrence River.* Canadian Technical Report of Fisheries and Aquatic Sciences No. 1876. Fisheries and Oceans.

INGRAM, R.G., and M.I. EL-SABH. 1990. Fronts and mesoscale features in the St. Lawrence Estuary. In *Oceanography of a Large-Scale Estuarine System: The St. Lawrence.* M.I. EL-SABH and N. SILVERBERG, eds. *Coastal and Estuarine Studies* 39:71-93. Springer-Verlag, Berlin.

JOHNSON, G. 1991. *Statistiques de pêche commerciale de 1986-1992 (mise à jour 1992).* Ministère de l'Agriculture, des Pêcheries et de l'Alimentation du Québec, Direction du développement et des activités régionales, Québec.

KENNEDY, J.A., and S. NADEAU. 1993. *La contamination de la sauvagine et de ses habitats par la grenaille de plomb au Canada.* Technical Report Series No. 164. Environment Canada, Canadian Wildlife Service, Ottawa.

KINGSLEY, M.C.S. 1995. Personal communication. Fisheries and Oceans, Maurice-Lamontagne Institute, Mont-Joli.

KINGSLEY, M.C.S. 1994. *Recensement, tendance et statut de la population de bélugas du Saint-Laurent en 1992.* Canadian Technical Report of Fisheries and Aquatic Sciences No. 1938. Fisheries and Oceans.

KINGSLEY, M.C.S., and M.O. HAMMILL. 1991. *Photographic census surveys of the St. Lawrence Beluga population, 1988 and 1990.* Canadian Technical Report of Fisheries and Aquatic Sciences No. 1776. Fisheries and Oceans.

KOUTITONSKY, V.G., and G.L. BUGDEN. 1991. The physical oceanography of the Gulf of St. Lawrence: A review with emphasis on the synoptic variability of the motion. In *The Gulf of St. Lawrence: Small Ocean or Big Estuary*? J.-C. THERRIAULT, ed. Proceedings of a symposium held at the Maurice-Lamontagne Institute, Mont-Joli, Quebec, March 14-17, 1989. Fisheries and Oceans. Canadian Special Publication of Fisheries and Aquatic Sciences No. 113. pp. 57-90.

LAPIERRE, L., and J. FONTAINE. 1994. *Distribution et abondance des Moules zébrées fixées sur les bouées de navigation illuminées du fleuve Saint-Laurent (1990, 1991 et 1992).* Environment Canada, Quebec Region, Environmental Conservation, St. Lawrence Centre, Montreal. Forthcoming.

LAVOIE, G. 1992. *Plantes vasculaires susceptibles d'être désignées menacées ou vulnérables au Québec.* Ministère de l'Environnement du Québec, Direction de la conservation et du patrimoine écologique, Québec.

LEGAULT, G., and M. VILLENEUVE. 1993. *Le CHIMIOTOX: Résultats d'évaluation chimio-toxique des établissements industriels du Plan d'action Saint-Laurent, Vol. 1.* Environment Canada and Ministère de l'Environnement du Québec, St. Lawrence Action Team, Montreal.

LEGENDRE, P., and H. SLOTERDIJK. 1988. *Synthèse de la contamination des poissons adultes du fleuve par le mercure, les BPC, le DDE, le mirex et les HCB.* Preliminary version. Environment Canada, Conservation and Protection, Quebec Region, St. Lawrence Centre, Montreal.

LEHOUX, D., A. BOURGET, P. DUPUIS, and J. ROSA. 1985. *La sauvagine dans le système du Saint-Laurent: Fleuve, estuaire, golfe.* Environment Canada, Conservation and Protection, Quebec Region, Canadian Wildlife Service, Québec.

LEMAY, A.-B., R. McNICOLL, and R. OUELLET. 1989. *Incidence de la grenaille de plomb dans les gésiers de canards, d'oies et de bernaches récoltées au Québec.* Ministère du Loisir, de la Chasse et de la Pêche, Direction de la gestion des espèces et de ses habitats, Québec.

LÉVEILLÉ, G. 1992. *Historique du classement des plages par région administrative et par municipalité, 1992.* Ministère de l'Environnement du Québec (1987, 1988, 1989, 1990 and 1991 editions also consulted).

LGL Ltd. (Le Groupe). 1990. *Inventaire des établissements industriels majeurs situés le long du fleuve Saint-Laurent et de la rivière Saguenay.* Prepared for Environment Canada and Ministère de l'Environnement du Québec, St. Lawrence Action Team, Montreal.

LORING, D.H., and D.J.G. NOTA. 1973. *Morphology and Sediments of the Gulf of St. Lawrence.* Bulletin of the Fisheries Research Board of Canada No. 182.

LORRAIN, S., and M. PELLETIER. 1993. Unpublished results of sediment contamination by trace metals and organic contaminants in northwestern Lake Saint-Louis. Environment Canada, Conservation and Protection, Quebec Region, St. Lawrence Centre, Montreal.

MAILHOT, Y. 1989. Personal communication. Ministère du Loisir, de la Chasse et de la Pêche, Région de Trois-Rivières.

MARQUIS, H., J. THERRIEN, P. BÉRUBÉ, G. SHOONER, and Y. VIGNEAULT. 1991. *Modifications physiques de l'habitat du poisson en amont de Montréal et en aval de Trois-Pistoles de 1945 à 1988 et effets sur les pêches commerciales.* Canadian Technical Report of Fisheries and Aquatic Sciences No. 1830. Fisheries and Oceans.

MENVIQ – MINISTÈRE DE L'ENVIRONNEMENT DU QUÉBEC. 1993. Unpublished data of the Banque de qualité du milieu aquatique (BQMA) databank, Québec.

MENVIQ – MINISTÈRE DE L'ENVIRONNEMENT DU QUÉBEC. 1992a. *Banque de données.* Centre de données sur le patrimoine naturel du Québec, Québec.

MENVIQ – MINISTÈRE DE L'ENVIRONNEMENT DU QUÉBEC. 1992b. *Programme d'assainissement des eaux du Québec (PAEQ).* 1992 computerized lists.

MENVIQ – MINISTÈRE DE L'ENVIRONNEMENT DU QUÉBEC. 1989. MUNFLEUV.DBF databank. Direction des écosystèmes urbains, Direction du réseau hydrique et Directions régionales.

MENVIQ and MSSS – MINISTÈRE DE L'ENVIRONNEMENT DU QUÉBEC and MINISTÈRE DE LA SANTÉ ET DES SERVICES SOCIAUX. 1993. *Guide de consommation du poisson de pêche sportive en eau douce.*

MINISTÈRE DU TOURISME DU QUÉBEC and REGIONAL TOURISM ASSOCIATIONS. 1991. Quebec tourism guides. Regions consulted: Montreal, Laval, Laurentians, Lanaudière, Montérégie, Coeur-du-Québec, Chaudière-Appalaches, Quebec City, Charlevoix, Saguenay–Lac-Saint-Jean, Bas-Saint-Laurent, Gaspésie, Manicouagan, Duplessis, Iles-de-la-Madeleine.

MLCP – MINISTÈRE DU LOISIR, DE LA CHASSE ET DE LA PÊCHE. 1987. *Nautisme Québec 1987 – Répertoire des marinas, quais pour petites embarcations et rampes de mise à l'eau.* Direction des communications, Québec.

MLCP – MINISTÈRE DU LOISIR, DE LA CHASSE ET DE LA PÊCHE. 1982. *Randonnée nautique 1982.*

MSSS and MENVIQ – MINISTÈRE DE LA SANTÉ ET DES SERVICES SOCIAUX and MINISTÈRE DE L'ENVIRONNEMENT DU QUÉBEC. 1992. *Guide de consommation du poisson de pêche sportive en eau douce.*

MTQ – MINISTÈRE DES TRANSPORTS DU QUÉBEC. 1991. *Carte routière du Québec 1990-1991.* Map (scale 1:500 000 and 1:1 000 000). Direction générale du Génie, Division de la cartographie, Québec.

OLIVIER, L., and J. BÉRUBÉ. 1993. *Qualité des sédiments et bilan des dragages sur le Saint-Laurent.* Environment Canada, Conservation and Protection, Quebec Region, St. Lawrence Centre, Montreal.

PARENT, M., and C. BOISVERT. 1979. *Programme des plages. Région de Montréal 1969-1978.* Bessette, Crevier, Parent, Tanguay and Associates for the Government of Québec, Services de Protection de l'environnement, Direction générale de la nature, Montreal.

PILOTE, S. 1993. Personal communication. Aquarium du Québec, Québec.

QUÉMERAIS, B. 1993. Personal communication. Environment Canada, Quebec Region, Conservation and Protection, St. Lawrence Centre, Ecotoxicology and Ecosystems Branch, Montreal.

REED, A. 1993. Personal communication. Environment Canada, Conservation and Protection, Quebec Region, Canadian Wildlife Service, Québec.

ROBITAILLE, J.-A., Y. VIGNEAULT, G. SHOONER, C. POMERLEAU, and Y. MAIHOT. 1988. *Modifications physiques de l'habitat du poisson dans le Saint-Laurent, de 1945 à 1984, et effets sur les pêches commerciales.* Canadian Technical Report of Fisheries and Aquatic Sciences No. 1608. Fisheries and Oceans.

RODRIGUE, J. 1993. Unpublished data. Environment Canada, Conservation and Protection, Quebec Region, Canadian Wildlife Service, Québec.

RUSCH, D.H., C. DAVIDSON ANKNEY, H. BOYD, J.R. LONGCORE, F. MONTALBANO III, J.K. RINGELMAN, and V.D. STOTTS. 1989. Population ecology and harvest of the American black duck: A review. *Wildlife Society Bulletin* 17(4): 379-406.

SAVARD, L., S. HURTUBISE and H. BOUCHARD. 1993. *Évaluation des agrégations de crevettes nordiques* (Pandalus borealis) *du nord du golfe du Saint-Laurent.* Research Document 93/20. Fisheries and Oceans, Maurice-Lamontagne Institute.

SERGEANT, D.E., and W. HOEK. 1988. An update of the status of White whales, *Delphinapterus leucas*, in the St. Lawrence Estuary, Canada. *Biological Conservation* 45:287-302.

SLC – ST. LAWRENCE CENTRE. 1993a. Unpublished data. Environment Canada, Conservation and Protection, Quebec Region, Ecotoxicology and Ecosystems Branch, Montreal.

SLC – ST. LAWRENCE CENTRE. 1993b. *The River at a Glance.* Info-flash report on the state of the St. Lawrence River, "St. Lawrence UPDATE" series. Environment Canada, Conservation and Protection, Quebec Region, Montreal.

SLC and LAVAL UNIVERSITY – ST. LAWRENCE CENTRE and LAVAL UNIVERSITY. 1991. *A River, Estuaries and a Gulf: Broad Hydrographic Divisions of the St. Lawrence.* Environmental Atlas of the St. Lawrence, "St. Lawrence UPDATE" series. Environment Canada, Conservation and Protection, Quebec Region, Montreal.

STATISTICS CANADA. 1993. *Le transport maritime au Canada – 1992.* Catalogue No. 54-205.

STATISTICS CANADA. 1992. *Le transport maritime au Canada – 1991.* Catalogue No. 54-205.

STATISTICS CANADA. 1991. *Chiffres de population et des logements. Division de recensement et subdivision de recensement.* Catalogue No. 93-304.

THERRIAULT, M. 1993. Personal communication. Fisheries and Oceans.

TRÉPANIER, S., G. SHOONER, J. THERRIEN, and M. BRETON. 1989. *Étude socio-économique sur l'industrie d'observation des baleines au Québec (mise à jour 1984-1988).* Gilles Shooner and Associates for Fisheries and Oceans, Quebec Directorate.

VERRETTE, J.-L. 1990. *Délimitation des principales masses d'eau du Saint-Laurent (Beauharnois à Québec).* Les Consultants Hydriques for Environment Canada, Conservation and Protection, Quebec Region, St. Lawrence Centre, Montreal.

VILLENEUVE, M. 1994. Personal communication. Ministère de l'Environnement du Québec, St. Lawrence Action Team, Montreal.